WORLD BANK ENVIRONMENT PAPER NUMBER 16

The Global Environmental Benefits of Land Degradation Control on Agricultural Land

Global Overlays Program

Stefano Pagiola

The World Bank
Washington, D.C.

Environment Papers are published to communicate the latest results of the Bank's environmental work to the development community with the least possible delay. The typescript of this paper therefore has not been prepared in accordance with the procedures for formal printed texts, and the World Bank accepts no responsibility for errors. Some sources cited in this text may be informal documents that are not readily available.

Stefano Pagiola is an economist in the World Bank Environment Department.

Library of Congress Cataloging-in-Publication Data

Pagiola, Stefano.
    The global environmental benefits of land degradation control on
agricultural land : global overlays program / Stefano Pagiola.
       p.   cm. -- (World Bank environment paper : no. 16)
    Includes bibliographical references (p.).
    ISBN 0-8213-4421-8
    1. Environmental degradation. 2. Land degradation--Environmental
aspects. 3. Land capability for agriculture. 4. Sustainable
agriculture. I. Title. II. Series.
GE140.P34 1999
333.76' 15--dc21
                                      98-53728
                                         CIP

Contents

Figures

Tables

Boxes

Foreword

This paper was prepared as part of the World Bank's Global Overlay Program. The Global Overlays Program, launched by the World Bank in partnership with bilateral donors and NGOs, seeks to internalize global externalities into national environmental planning and the Bank's sector work, operations, and dialogue with governments and partners. It is an iterative process, combining conceptual studies, reviews of state-of-the-art techniques for measuring and mitigating global externalities, and testing these concepts and tools in country-level studies as a means of identifying good practices for country planners and Bank task managers. The results will help guide national actions to reduce greenhouse gas emissions, conserve biodiversity, and protect international waters.

Global overlays add a new dimension to traditional sector economic planning by analyzing environmental impacts and opportunities to internalize global externalities. This analysis asks: How and at what cost would policies, institutions, and investment priorities change if global environmental objectives were added to conventional sectoral objectives?

Acknowledgments

This paper was written by Stefano Pagiola of the World Bank's Environment Department, under the guidance of Lars Vidaeus, Chief of the Global Environment Division.

This paper draws heavily on the existing literature and documentation of lessons learned from Bank and other projects. Background papers on the linkages between land degradation and climate change and biodiversity were prepared by Susan Leloup (consultant).

Useful comments were provided by Graeme Donovan and Jeff Lewis (Africa Region); Robin Broadfield, Noreen Beg, Kathy McKinnon, Ernst Lutz, and Natsuko Toba (Environment Department); Julian Dumanski and Christian Pieri (Rural Development Department); and Ken King and Kanta Kumari (GEF Secretariat).

This paper has also benefitted from discussions with Jan Bojö, Robert Clement-Jones, Jumana Farah, Isabel Valencia, Francisco Pichon, Shiv Singh, and Bachir Souhlal (Africa Region); Douglas Graham and John Kellenberg (Latin American and Caribbean Region); John Dixon, Kristin Elliott, Sam Fankhauser, Hassan Hassan, and Robert Watson (Environment Department); Doug Forno and Bill Magrath (Rural Development Department); John English (Operations Evaluation Department); Walter Lusigi (GEF Secretariat); Stein Hansen (GEF Scientific and Technical Advisory Panel); Nessim Ahmad, Willem Bettink, and Lorenz Petersen (IFAD); Timo Maukonen (UNEP); Pedro Sanchez and Meredith Soule (ICRAF); and Leonard Berry (Florida Atlantic University).

Executive Summary

Land degradation on agricultural land threatens the sustainability of growth and the welfare of the many people who depend on agriculture for their livelihoods—including many of the poorest members of the world's population. It can also have adverse effects on problems of global significance, including climate change, biodiversity, and international waters. In some cases, there may be important complementarities between measures that address the national and the global aspects of land degradation problems.

The Convention to Combat Desertification (CCD) defines land degradation as a "reduction or loss ... of the biological or economic productivity and complexity of rainfed cropland, irrigated cropland, or range, pasture, forest and woodlands" (art.1). This note focuses on land degradation in areas used for agricultural production, including croplands and rangelands, and focuses narrowly on instances in which the productive potential of the land is reduced as a result of land use practices. The effects of land use change in and of itself (including deforestation) will not be considered.

Land degradation can cause problems at three levels:
- At the *field level*, land degradation can result in reduced productivity.
- At the *national level*, land degradation can cause problems such as flooding and sedimentation.
- At the *global level*, land degradation can contribute to climate change, and to damage to biodiversity and international waters.

The relative importance, and specific nature, of problems at each level varies substantially from case to case. Although this note focuses on the global effects of land degradation, it is important to remember that local and national effects are usually the most important.

Significant gaps remain in our understanding of land degradation. Even at the field and national levels, where problems are generally well understood qualitatively, quantitative data are often insufficient to allow specific problems to be analyzed. At the global level, the weakness of the quantitative data is compounded by a much poorer qualitative understanding.

Effects on Climate Change

Emissions. Terrestrial ecosystems are an important carbon sink. The stock of carbon stored in the upper 1m of the world's soils is estimated to be about 1.5 times the amount of carbon stored in biomass. Soil carbon is lower on agricultural lands, but the amount stored is nevertheless much higher than might be expected from above-ground biomass alone. However, major uncertainties remain with regard to the role of the terrestrial carbon pool.

The question of interest here is the extent to which land degradation on agricultural land affects climate change.
- Does land degradation on agricultural land result in increased emissions of greenhouse gases?
- Does land degradation on agricultural land affect it's capacity to serve as a carbon sink?
- Can appropriate management enhance both land's productivity and its capacity to store carbon?

Unfortunately, data to answer these questions are scarce. Most attention has focused on emissions under specific land uses, and on the impact of land use change (in particular, on the impact of deforestation). Very little work has been done on the effect of land degradation within a given land use on emissions.

Carbon cycle in soils. Because croplands and rangelands tend to have relatively low above-ground biomass, the linkages between land degradation and climate change are likely to come primarily from changes in soil carbon.

• Some actions which *cause* land degradation can increase carbon emissions directly. For example, burning crop residues causes both fertility loss, by preventing the return of nutrients to the soil and reducing the build-up of soil organic matter, and increased emissions.

• Some forms of degradation reduce soil carbon. For example, erosion can carry away soil organic matter. However, this does not always lead to increased emissions since carbon carried away by erosion is often deposited under conditions where it may be well preserved, such as in reservoirs.

• The *consequences* of land degradation also adversely affect the soil carbon cycle. Lower production of crops and pasture, results in lower carbon inputs in subsequent periods (less root material, less leaf litter, less crop residue), thus reducing carbon storage.

Land degradation on croplands and rangelands is thus likely to reduce the ability of soils to serve as a carbon sink and release carbon currently stored in soils to the atmosphere. The magnitude of this effect is difficult to estimate, however.

Effects on Biodiversity

Croplands. Croplands are substantially modified from their original, natural state, and their levels of biodiversity are generally substantially lower than those of natural habitats. Nevertheless, agricultural landscapes can contain considerable biodiversity. The concern here is whether land degradation might further reduce the remaining biodiversity.

Below-ground biodiversity. The main direct adverse effect of cropland degradation on biodiversity is likely to be on below-ground biodiversity. Diverse and abundant organisms help maintain soil fertility and productivity and are fundamental to soil quality. Degradation of soil physical and chemical conditions can damage this biodiversity, about which relatively little is known.

Indirect effects. In most cases, the greatest impact of cropland degradation on biodiversity is likely to be indirect. By reducing productivity on existing agricultural land, degradation can force farmers to clear additional areas of natural habitat to maintain production. It should be remembered, however, that land degradation is not the only cause of agricultural expansion.

Rangeland. Rangelands tend to be less modified from their natural state than cropland and to contain a much greater proportion of their original biodiversity. Livestock often shares rangelands with considerable wildlife. Degradation, therefore, can cause relatively more damage to biodiversity on rangelands than on croplands.

Many of the factors that cause pasture degradation are also likely to have an adverse impact on biodiversity. For example, both livestock and wildlife will suffer if access to areas that provide critical grazing or water at times of stress are restricted. As both livestock and wildlife are restricted to smaller, often less favorable areas, competition between them is likely to be exacerbated.

Effects on International Waters

Many off-site consequences of land degradation may be experienced beyond national borders. Sedimentation or flooding problems caused by degradation in an upstream watershed, for example, may affect a country downstream. Damage to international waters are a special case of the off-site damages. They are similar to off-site effects felt nationally, except that national policymakers have no incentives to take them into account. The problems of identifying and measuring cause-and-effect relationships are particularly severe in the case of international waters, since data collected in upstream and downstream countries may not be compatible.

Global Benefits of Land Degradation Control

Measures to control land degradation also have effects at the field, national, and global levels. The specific range of benefits obtained depends on the measures used and the conditions under which they are applied.

Land degradation control can help reduce or halt the adverse global effects of land degradation. Some land degradation control practices can also have global benefits in and of themselves, by stimulating additional carbon sequestration and/or biodiversity conservation over and above what might have occurred even in the absence of degradation. Practices likely to have positive effects on problems of global concern include *agroforestry*, which allows increased carbon sequestration while continuing with crop production and which often provides a more hospitable environment for biodiversity, and *community-based wildlife management*, which can provide an alternative to unsustainable use of some marginal areas.

Integrating Global Dimensions into Land Degradation Control Projects

The primary reason for efforts to control land degradation on agricultural land is to reduce, arrest, or reverse the field-level or national problems it is causing. Given the linkages between land degradation and problems of global concern, however, global benefits may also be generated in some cases.

When linkages between land degradation and problems of global concern exist, efforts to control and reverse them can be mutually supportive. Land degradation control activities can be win-win in the sense that they reduce both the local and the global effects of land degradation. In some cases, however, addi-

tional or different measures may be required to fully realize global benefits.

Any effort to incorporate global dimensions into land degradation control must begin with a well-thought out strategy to address the local and national aspects of the problem. This requires a clear understanding of the nature, extent, and severity of land degradation problems, their causes, and their effects at both the farm and national levels. It also requires a clear understanding of the incentives and constraints faced by land users.

At this stage, it should be possible to determine whether the measures already envisaged are sufficient to address the global problems originating at the site, or whether additional or different measures are required to do so. In such cases, it may be possible to obtain funding from the Global Environment Facility (GEF) to finance the incremental costs of the additional measures. It is important to realize, however, that not all activities that generate global benefits are eligible for GEF funding. The GEF's Operational Programs set specific priorities as well as specifying eligibility criteria.

The main difficulties likely to be encountered by efforts to incorporate global considerations in land degradation control activities are:
- *Information.* Information on the nature and magnitude of the adverse global effects of land degradation under specific conditions is extremely scarce.
- *Implementation.* To be successful, land degradation control programs need to obtain the cooperation of land users. Insufficient attention to the constraints and incentives land users face has led to the failure of many land degradation control projects.

Pilot Project Concepts

In view of the limited experience in the preparation of projects that blend global environmental concerns with land degradation control, the World Bank and the International Fund for Agricultural Development (IFAD) have collaborated in developing a pipeline of projects in this nascent area of GEF operations. Project concepts were developed for Botswana, Mali, Jordan, India (2 projects), Mongolia, Belize, and El Salvador. Although these pilot project concepts are only illustrative of the potentials and pitfalls of attempting to integrate attention to global benefits into land degradation control projects, they do provide some initial lessons.

The projects which seem to lend themselves best to integrating global dimensions are those in which field activities are being carried out in specified areas. It is far more difficult to identify potential global dimensions of projects that seek to combat land degradation through policy reforms or support to research and extension. Conversely, when field activities are being carried out, identifying possible links to problems of global concern is substantially easier—particularly in the case of links to biodiversity.

The experience of the pilot projects also suggests that establishing the nature and extent of linkages between land degradation on agricultural land and biodiversity is simpler than doing so for climate change. In the case of biodiversity, the main constraint is that the biodiversity at risk from land degradation on agricultural land may not have been as well studied as biodiversity in protected areas. In the case of climate change, the main constraint is that very few data exist on changes in emission resulting from changes *within* a given land use (as opposed to changes in emissions resulting from changes in land use itself, such as deforestation). The data requirements are also more stringent in the case of climate change, due to the need to demonstrate cost-effectiveness.

The Role of the World Bank

The World Bank is devoting considerable resources to assist its client countries to combat land degradation. The World Bank is also committed to addressing global environmental problems. Through its Global Overlays Program, the World Bank is seeking to incorporate attention to global problems throughout its work. This includes seeking more systematically to assess the impacts on biodiversity and other global externalities that land degradation and its control might generate, and incorporating consideration of these issues in policy and project responses.

Conclusion

For many countries—and in particular for many African countries—land degradation on agricultural land is posing substantial threats to sustainability, economic growth, and the welfare of the rural population. Strong efforts to combat land degradation are justified on these grounds alone. In some cases, reduction of problems of global concern such as mitigation of climate change or conservation of biodiversity provide an additional reason to combat degradation. At times this may require additional or different measures than if local and national considerations were the only ones involved.

In cases where there are strong linkages between land degradation and problems of global concern, efforts to combat both can be mutually supportive. It is important, however, to remember that the primary motivation for land degradation control efforts will remain the local and national benefits that can be derived thereby. Linkages to global problems are not always present, or may not be sufficiently strong to warrant specific attention.

Acronyms and abbreviations

ASR   Agricultural Sector Review
BSAP   Biodiversity Strategy and Action Plan
CBNRM Community-Based Natural Resource Management
CBD   Convention on Biological Diversity
CCD   United Nations Convention to Combat Desertification
CESP   Country Environmental Strategy Paper
CGIAR   Consultative Group on International Agricultural Research
CH_4 methane
CO_2 carbon dioxide
EIA   Environmental Impact Assessment
ESW   Economic and Sector Work
FAO   Food and Agriculture Organization of the United Nations
FCC   United Nations Framework Convention on Climate Change
GEF   Global Environment Facility
GHG   greenhouse gas
IFAD   International Fund for Agricultural Development
LQI   Land Quality Indicators
NAP   National Action Program
NARS   National Agricultural Research System
NEAP   National Environmental Action Plan
NGO   non governmental organization
N_2O nitrous oxide
OP   Operational Program
PDF   Project Development Facility
SFI   Soil Fertility Initiative
UNDP   United Nations Development Programme
UNEP   United Nations Environment Programme
USDA   United States Department of Agriculture

ha   hectares
Mg   megagram (1×10^6 grams = 1 metric tonne)
Pg   petagram (1×10^{15} grams = 1 billion metric tonnes)

All monetary amounts are in U.S. dollars.

1. Introduction

Since 1945, an estimated two billion hectares of agricultural land—almost 18 percent of the earth's vegetated land—have been degraded as a result of human activity [Oldeman and others, 1990]. Of these, an estimated 1.2 billion hectares, or almost 11 percent of the earth's vegetated land, have been moderately or strongly degraded, implying that productivity has been significantly reduced (Figure 1). Land degradation poses a substantial threat to the sustainability of development [World Bank, 1992]. In the drylands of Sub-Saharan Africa in particular, land degradation is thought to be an important impediment to agricultural growth [Cleaver and Schreiber, 1994].

Controlling land degradation on agricultural land is important to the objectives of sustainable growth and increasing the welfare of the

Figure 1
Estimates of human-induced land degradation by region, 1945-1990

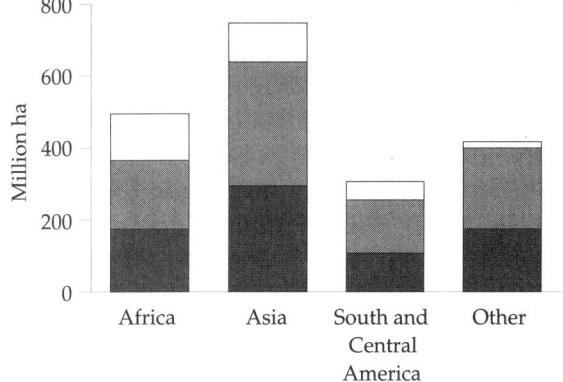

Strong degradation: Agricultural use impossible
Moderate degradation: Significant productivity decline
Light degradation: Small productivity decline
Source: From data in Oldeman and others, 1990

many people who depend on agriculture for their livelihoods—including many of the poorest members of the world's population. The urgency of addressing land degradation problems has been reiterated by the adoption and ratification of the Convention to Combat Desertification (CCD), which came into force in December 1996. Under this convention, many countries have committed themselves to combating land degradation (Box 1).

Although many of the consequences of land degradation are experienced either in the specific areas being degraded or in neighboring areas, it is increasingly thought that land degradation can also have adverse effects on problems of global significance:

- *Climate change.* Land degradation can contribute to global climate change by leading to emission of greenhouse gases (GHGs) or by reducing the ability of terrestrial ecosystems to act as a sink for these gases.
- *Biodiversity.* Land degradation can result in loss of biodiversity, both in the areas degraded and by inducing additional clearing of natural habitats.
- *International waters.* Land degradation can result in damages to shared international water bodies.

In many cases, there are likely to be important complementarities between measures that address the national and the global aspects of land degradation problems.

Under the terms of the Instrument Establishing the Restructured Global Environment Facility (GEF), the agreed incremental costs of activities concerning land degradation, primarily desertification and deforestation, as they relate to the four focal areas of the GEF (climate

1

Box 1. The Convention to Combat Desertification

The United Nations Convention to Combat Desertification (CCD) became effective in December 1996, following its ratification by 50 signatories. To date, 150 countries have ratified the CCD.

The objectives of the Convention are to combat desertification (defined as land degradation occurring in arid, semi-arid, and dry sub-humid areas) and mitigate the effects of drought in countries that experience them. The Convention gives priority to Africa while not neglecting other regions (art.7).

Parties to the convention commit themselves to adopting an integrated approach to addressing the physical, biological, and socioeconomic aspects of desertification and drought. Efforts to combat desertification are to be integrated with strategies for poverty eradication. To meet the objectives of the CCD, parties are to be guided by three principles:
• stakeholder participation;
• international cooperation; and
• consideration of the specific needs of affected developing countries.

Affected parties are to develop National Action Programs (NAPs) which will
• define and promote preventive measures;
• enhance climatologic, meteorological, and hydrological capabilities;
• strengthen institutional frameworks;
• provide for effective stakeholder participation; and
• review implementation regularly.

Subregional and Regional Action Programs (SRAPs), which have the same basic features, will support implementation at a regional level.

The Convention will not have any financing of its own. However, it establishes a Global Mechanism (GM) which aims to improve management, mobilization, and coordination of funds for combating desertification. Developed country parties are committed to providing assistance to affected countries to implement their NAPs, to provide financing and other forms of support, and to mobilize new and additional funding from the GEF for the agreed incremental costs of activities concerning desertification that relate to the GEF's four focal areas.

change, biodiversity, ozone depletion, and international waters), are eligible for funding (art.I, para 2). Land degradation control is also integrated in the GEF Operational Strategy [GEF, 1996a; 1997].

The World Bank is committed, through its environmental agenda, to addressing global environmental challenges. As one of the implementing agencies for the GEF, the World Bank also has a direct responsibility to help its client governments to address problems of global concern. Although land degradation control has the potential of contributing to alleviation of global problems in these areas, there is little experience with integrating global considerations into land degradation control activities.

Definitions. The term 'land degradation' has been used to describe a wide variety of problems. This study uses the definitions in the Convention to Combat Desertification (art.1):
• *Land* means the terrestrial bio-productive system that comprises soil, vegetation, other biota, and the ecological and hydrological processes that operate within the system; and
• *Land degradation* means reduction or loss ... of the biological or economic productivity and complexity of rainfed cropland, irrigated cropland, or range, pasture, forest and woodlands resulting from land uses or from a process or combination of processes, including processes arising from human activities and habitation patterns, such as (i) soil erosion caused by wind and/or water; (ii) deterioration of the physical, chemical and biological or economic properties of soil; and (iii) long-term loss of natural vegetation.

For the purpose of this study, these definitions are interpreted narrowly to focus on instances in which the productive potential of the underlying land is reduced as a result of land use practices. Land use change in and of itself will not be considered land degradation. Only if the new use results in damage to the land's productive potential—and hence results in either falling yields or in the need for higher input levels to maintain yields—will it fall within the scope of this study.

Land degradation problems occur in forest land, in cropland, and in rangeland. This paper focuses on problems encountered in cropland

and rangeland—the areas used for agricultural production. A separate Global Overlay study addresses deforestation and forest management problems [Kellenberg and Cassells, forthcoming]. Within agricultural land, only effects related to land degradation are considered. Many aspects of agricultural production can affect problems of global concern even in the absence of degradation. A previous Global Overlays study [Pagiola and others, 1997] has already examined the interactions between agricultural development and biodiversity.

Land degradation can cause problems at three levels:

- At the *field level*, land degradation results in reduced productivity.
- At the *national level*, land degradation results in a range of problems such as damage to downstream infrastructure through flooding and sedimentation, reductions in water quality, and changes in the timing and quantity of water flows.
- At the *global level*, land degradation can contribute to climate change (through increased emissions of greenhouse gases and changes in the ability of terrestrial ecosystems to serve as carbon sinks), to damage to biodiversity (both directly in degraded areas and indirectly by inducing expansion of cultivated areas), and to damage to international waters (through sediment loads and changes in hydrological cycles).

The relative importance, and specific nature, of problems at each level varies substantially from case to case.

Conceptual framework. The conceptual framework used in this paper is based on a number of propositions that are embedded in Figure 2. Most land resource management decisions are made by individual land users such as farmers and pastoralists, not by national planners. The incentive structure under which land users make land resource management decisions is influenced by the technology they have available, by the structure of markets they operate in (including markets for the land itself), by the prices they face for their inputs and outputs

---

Box 2. Desertification

The term 'desertification' is avoided in this study, for two main reasons. First, it has been subject to a myriad of definitions over the years, many of them vague and imprecise. It is not, therefore, a useful term from either an analytical or a descriptive perspective [Nelson, 1990]. Second, considerable controversy has developed over whether any phenomenon which might reasonably be described as desertification even exists. Analysis of vegetation change in the Sahara by NASA, for example, clearly shows that vegetation fluctuates significantly [Tucker and others, 1991]. The current northern limits of vegetation are very similar to those preceding the severe droughts of the 1970s and 1980s. The image of advancing sand dunes which gave desertification its emotional impact, therefore, is a poor representation of reality.

Even the Convention to Combat Desertification, despite its name, moves away from the term by defining desertification as "land degradation in arid, semi-arid and dry sub-humid areas" (art. 1).

(including the opportunity cost of their family labor), and by the characteristics of the farm household. In turn, many of these elements are influenced, to varying degrees, by agriculture *and* non-agriculture policies, institutions, and development programs.

The resulting land use practices will, of course, affect the level of production, whether that production be a crop or pasture. Some land use practices will also affect the quality of the land; in particular, some land use practices can lead to land degradation. For example, growing crops that leave the soil exposed can result in erosion, and over-grazing can result in compaction of pasture lands and in a change in species composition. Some—often most—of the effects of degradation are felt *on-site*, at the field level. To the extent that they are, they will affect the future benefits that farmers can obtain from the land. Land-users, therefore, generally have a direct incentive to take on-site effects into account in their management decisions.

Degradation can also have *off-site* effects such as sedimentation. Land users generally do not have any incentive to take these effects into account. Since these problems affect national

Figure 2
Conceptual framework

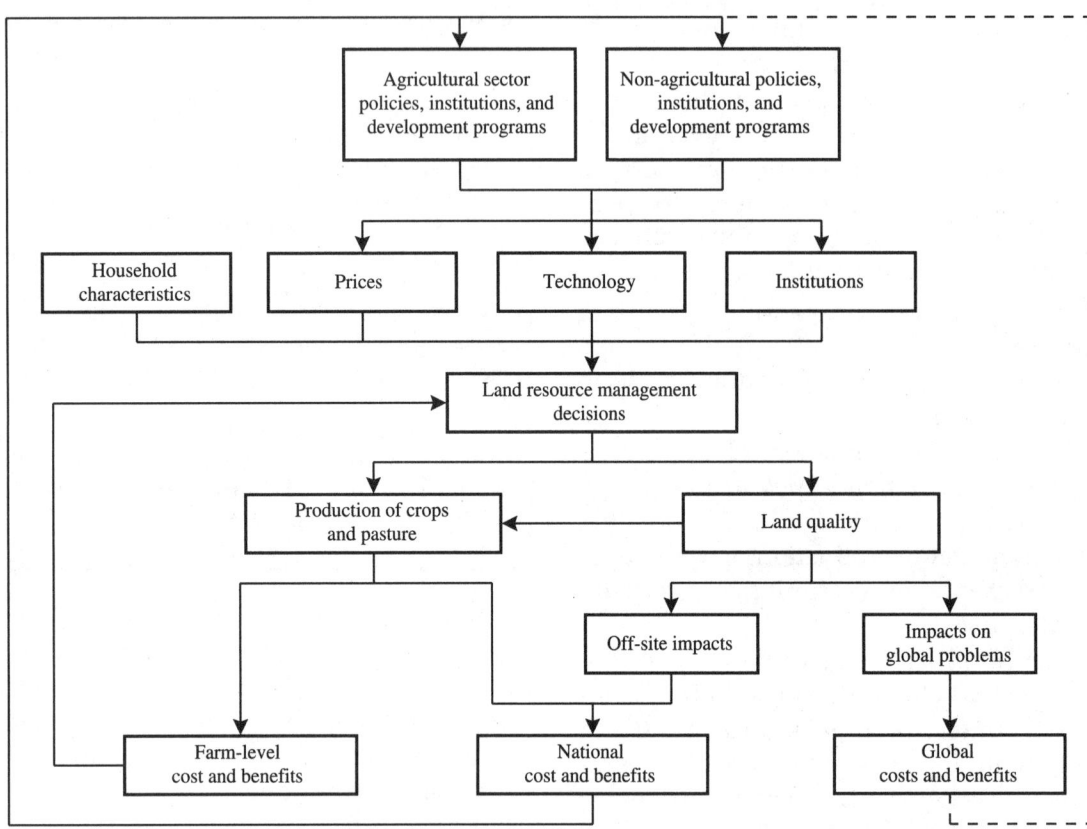

costs and benefits, national planners have an incentive to modify policies or design project interventions to attempt to direct farmers' land resource management decisions in socially optimal ways.

Land quality changes can also affect problems of global concern, through their effect on carbon sequestration, on-site and off-site biodiversity, and international waters. In the past, there has been no feedback from these problems, and so they have been ignored both by farmers and by national planners.

This study focuses on the missing link in Figure 2, indicated by the dotted line: on measures to address the global dimensions of land degradation and on how to integrate them with measures designed to address farm-level and national-level problems. As with other stu-

dies in the Global Overlay Program, this study asks the question: How and at what cost would policies, institutions, and investment priorities change if global environmental objectives were added to conventional sectoral objectives? The basic aspects of land degradation problems at the field and national levels are relatively well understood, at least in a qualitative sense. There is also a substantial body of work on how to design projects that address land degradation problems at these levels. In contrast, important gaps remain in our understanding of the global dimensions of land degradation. Moreover, there is little experience in preparing land degradation control projects that blend global environmental concerns with national sustainable development objectives.

Key issues. This paper examines the conceptual, methodological, and empirical issues involved in designing interventions to address land degradation problems and global environmental benefits in developing countries. The issues to be addressed include:

- The extent and nature of global benefits resulting from control and mitigation of land degradation problems.
- The relationship between global and domestic benefits of addressing land degradation problems.
- How consideration of global problems can be incorporated in the design of land degradation control projects and how the costs of such projects might be shared among the beneficiaries.

Chapter 2 discusses the global dimensions of land degradation from a technical viewpoint. The chapter begins with a brief overview of the on-farm and national effects of land degradation, before examining the relationship of land degradation to problems of global concern —climate change, biodiversity, and international waters. As will be seen, significant gaps remain in our understanding of land degradation problems. Even at the field and national levels, where problems are generally well understood qualitatively, quantitative data are in most instances insufficient to allow specific problems to be analyzed. At the global level, the weakness of the quantitative data is compounded by a much poorer qualitative understanding. The chapter then discusses in what ways land degradation control efforts might generate global benefits, and examines some measures which appear particularly likely to generate global benefits.

Chapter 3 discusses land degradation control programs and how attention to global problems might be incorporated into them. The chapter begins with a discussion of the incentives and constraints faced by land users. It then reviews previous efforts to address land degradation problems and the lessons which have been learned from these efforts. If land degradation control efforts are modified to 'go the extra mile' to generate additional global benefits, it may require land users and national authorities to undertake actions which do not bring direct local or national benefits. This may require external compensation from agencies such as the Global Environment Facility (GEF). The chapter then reviews a number of pilot project concepts prepared by the International Fund for Agricultural Development (IFAD) in collaboration with the World Bank that integrate attention to attention to global problems in land degradation control projects.

Chapter 4 summarizes the discussions and indicates the next steps required to advance the process.

Much of the discussion in this paper is broadly applicable to land degradation problems worldwide. However, the emphasis is primarily on conditions encountered in the drylands of Sub-Saharan Africa—the area where land degradation problems are thought to be most pressing. This emphasis is consistent with the CCD, which gives priority to Africa while not neglecting other regions (art.7).

This paper should not be interpreted as a good practice handbook on incorporating global dimensions in land degradation control. Too little work has been done in this field to date, and too many gaps remain in our knowledge of the effects of land degradation on global problems. Rather, this paper is a first step in an intellectual journey that should ultimately lead to good practice. Because land degradation problems are so diverse, many generalizations and an abundant use of qualifiers are inevitable.

2. Global Dimensions of Land Degradation on Agricultural Land

Land degradation on agricultural land causes a wide array of problems, at many different levels. Problems at the field and national level have been studied extensively and, although quantitative data are often missing, are relatively well understood. The global dimensions have been much less studied, and our understanding is weak even at a qualitative level.

This chapter begins with an overview of the field-level and national-level consequences of land degradation problems. Because the different aspects of land degradation problems are closely intertwined, any efforts to address the global dimensions of land degradation require an understanding of its local effects as well. The chapter then describes current knowledge on the possible effects of land degradation on problems of global concern: climate change, biodiversity conservation, and international waters.

Overview of Land Degradation Problems

Agriculture is dependent upon natural biophysical environments; degradation of these environments through misuse or over-use has led to considerable concern that agricultural production may not be sustainable [World Bank, 1992; World Resources Institute, 1992]. A global assessment commissioned by the United Nations Environment Programme (UNEP), found that almost 11 percent of the earth's vegetated land has been moderately or strongly degraded, implying that productivity has been significantly reduced [Oldeman and others, 1990]. The extent of degradation is estimated to be particularly high in Africa, where about 320 million ha are moderately or strongly degraded (see Figure 1 on page 1 above). Stoorvogel and Smaling [1990] estimate that, over the last 30 years, an average of 660 kg/ha of nitrogen, 75 kg/ha of phosphorus, and 450 kg/ha of potassium have been lost from cultivated land in Sub-Saharan Africa, primarily because of insufficient use of nutrient inputs to offset nutrient output through crop removals and other losses.

Considerable debate has arisen, however, over the magnitude of the problem. Recent reviews of available evidence on land degradation in several countries challenge the more catastrophist views of its nature, extent, and severity (Box 3). What emerges in these and other studies is a much more nuanced picture of land degradation, in which site-specific problems dominate and in which a complex series of causes lead to degradation.

Data limitations. Any effort to examine land degradation problems soon runs into formidable data limitations. Despite decades of work on land degradation, the available information remains highly fragmented, incomplete, and often unreliable. In particular, much of the research on land degradation stops short of addressing the productivity issues which are of fundamental interest. Few of the many studies which examine erosion problems, for example, go beyond measuring erosion in tons of soil lost per hectare and also measure the resulting effect on crop production. Although a range of efforts are underway to remedy these problems (such as the Land Quality Indicators Program, see Box 9 on page 39 below), the gradual and

---

Box 3. How severe is land degradation? Country-level evidence

Recent reviews of available evidence on land degradation in several countries challenge the more catastrophist views of its nature, extent, and severity.

- *Kenya*. A review of the experience of the Machakos district in Kenya, where land degradation problems have been widely noted beginning in the 1930s, has shown that increasing population pressure need not lead to increased degradation. Despite a sixfold increase in population, productivity increased both in per hectare and per capita terms, and degradation was controlled and even reversed [English and others, 1992; Tiffen and others, 1994].
- *El Salvador*. Pagiola and Dixon [1997] examined available data on land degradation in El Salvador, where the common perception is that '75% of the country's surface is degraded', and found that only about a third of farmers' fields experience erosion, and only a fraction of those appear likely to suffer productivity declines as a result.

None of these studies suggests that land degradation is *not* a problem. All, however, seriously question the common perception of the severity and nature of land degradation.

---

cumulative nature of land degradation means any effort to collect improved data will necessarily take time.

On-site Problems

The on-site effects of land degradation on agricultural land are a major source of concern, since they threaten the sustainability of agricultural production and the welfare of a substantial portion of the world's population.

Cropland

Land degradation on cropland can take a variety of forms. Traditionally, erosion and its consequences have attracted the most attention [Brown and Wolf, 1986]. More recently, interest has focused on the effects of nutrient depletion. The fertility constraints resulting from soil organic matter and nutrient depletion are thought to be a major impediment to agricultural

growth in Sub-Saharan Africa [Woomer and Swift, 1994; Sanchez and others, 1996b, 1997]. Other important forms of degradation include salinization, which is an important threat to irrigated areas in semi-arid regions [Umali, 1993]. In many cases, different forms of degradation are correlated. For example, soil compaction can result in increased runoff and, hence, in higher rates of erosion; conversely, erosion can carry away nutrients and weaken the soil's physical structure. An important characteristic of such damage is that it is usually cumulative; its effects in any one year can be minor or insignificant, but become important as they accumulate over time [Lal, 1987].

Despite years of concern over the effects of land degradation on cropland, critical gaps remain in available data. In particular, much of the data are purely qualitative. Land is variously classified as 'undegraded', 'degraded', or 'severely degraded', but the meanings of these labels are seldom made precise. Data on the productivity effects of degradation are particularly scarce.

Examination of available data on crop yields do not provide strong evidence that degradation is affecting productivity. On the contrary, yields appear to have increased in many countries, including many in Sub-Saharan Africa. (In much of Africa, however, this increase has been slower than population growth, so that *per capita* production has fallen.) Taken alone, however, these data do not disprove the existence of degradation.

- National yield data are often suspect, both in terms of data quality and in terms of representativeness. If data collection tends to favor more prosperous farmers using improved techniques, for example, it may miss degradation problems experienced by the bulk of farmers.
- Both agricultural technology and input levels have been increasing (although again, more slowly in Sub-Saharan Africa than in other regions), and these improvements may have offset some of the effects of degra-

dation. In the absence of degradation, yield increases might have been faster.

- Expansion of agriculture into new, as yet undegraded, areas may mask the effects of degradation on existing agricultural land. Continued expansion will be increasingly difficult, however, and will bring into use increasingly marginal land.
- Data on average national yields may well conceal significant regional variations.

Although efforts are clearly needed to improve available data on the productivity effects of land degradation, it would be unwise to assume from the current weak evidence that such problems do not exist.

Rangelands

To an even greater extent than cropland, rangeland—particularly in Africa—has been perceived as suffering from widespread degradation. Indeed, conditions in semi-arid and arid rangelands played a major role in the popular perception of desertification. Dregne and others [1992], for example, estimate that the majority of the world's rangelands are moderately or severely degraded. The main problem was diagnosed as being overstocking of livestock by pastoralists, resulting in overgrazing which removed vegetation, changed species composition towards unpalatable species, and compacted soils. This was thought to result in a collapse—perhaps irreversible—of rangeland productivity, drastically reducing its carrying capacity.

As with cropland, however, there is a growing realization that this image may be misleading. Recent research has argued that the extent of dryland degradation is much lower than has been believed and that rangelands are often more resilient than had been thought. Contrary to conventional wisdom, animal production per hectare and meat production per head in the Sahel has increased over the last 30 years [Steinfeld and others, 1997]. The diagnosis of the problem has also been changing substantially [Behnke and Scoones, 1993].

Rangelands in arid and semi-arid systems are now recognized as being highly variable over both space and time (so that much apparent degradation was actually the temporary effect of low rainfall). Their productivity at any given time is driven primarily by factors such as rainfall, rather than by stocking density. Moreover, considerable evidence now demonstrates that traditional pastoral strategies, which rely on extensive movement in response to stress, are not only well-suited to their environment, but can also outperform 'modern' ranching strategies [Scoones, 1995].

Here too, a realization that conventional wisdom may have exaggerated and mis-diagnosed problems does not imply that rangeland degradation does not occur. The causes of degradation, however, lie more in encroachment into grazing land by settled agriculture, in obstacles to movement by livestock in response to climatic variations, and in the breakdown of traditional communal management arrangements than in excessive livestock numbers. Government subsidies during times of low rainfall, although they may alleviate short-term hardship, may cause long-term problems by maintaining high stocking rates and preventing ecologically normal regeneration of vegetation after drought [Pratt and others, 1997].

Off-site Problems

In addition to affecting the productivity of the land directly affected (on-site effects), some forms of degradation can also cause damages elsewhere (off-site effects). In particular, erosion can cause economic damage to reservoirs and waterways and to aquatic life within them. In some cases, off-site effects can be much more important than on-site effects. In the USA, for example, changes in soil productivity have been estimated to be relatively minor compared to off-site costs [Crosson and Stout, 1983; Clark and others, 1985]. In developing countries, on the other hand, on-site productivity concerns tend to be dominant. Magrath and Arens [1989], for example, estimate that productivity effects account for 95 percent of the costs of soil erosion in Java (although several categories of off-site costs could not be quantified). Repetto and Cruz [1991] obtain similar results in Costa Rica, albeit from weak and incomplete data.

Sedimentation. The off-site effect which has led to the most concern is the sedimentation of reservoirs and waterways resulting from upstream erosion. In Morocco, for example, it is estimated that up to 0.5 percent of national water storage capacity is lost annually to sedimentation, a large amount for a country with rising water shortages [Agro-Concept, 1994]. Sedimentation can have a number of adverse consequences:

- Higher costs of hydroelectricity production and shorter reservoir lifetimes;
- Damage to irrigation systems (reductions in irrigated area, higher costs of canal cleaning, damage to pumping equipment);
- Reduced ability to regulate streamflow, increasing the danger of flooding;
- Damage to fish stocks and other aquatic lifeforms through increased water turbidity and burial of spawning grounds and coral reefs.

These costs can be avoided by dredging reservoirs and waterways to remove sediment or by building new infrastructure to replace that which has been silted up, but this is costly.

However, the contribution of anthropogenic land degradation to sedimentation problems is often not known. In many cases, it is implicitly assumed that all sedimentation is anthropogenic. This ignores the often very high rates of background erosion in many areas. In Morocco, for example, the bulk of sediment is now thought to originate from streambank erosion and not from erosion on agricultural land. Even among anthropogenic sources, agricultural land may not be the most important. On a per hectare basis, roads are often the single biggest source of sediment. Even when cropland and rangeland contribute to sedimentation, it is difficult to establish the contribution of specific fields. Most soil eroded from a given field is simply re-deposited nearby.

Box 4. Site-specificity of land degradation

An important aspect of land degradation problems, at all levels, is their site-specificity.

Field-level problems. Soil characteristics vary widely from place to place, depending on climate, parent material, topography, biotic activity, and the length of time that soil formation has been underway. In the United States alone, 13,000 distinct soil series have been identified [Brady, 1986]. Climatic variations and crop interactions add additional layers of complexity. Site-specific variations in conditions can lead to significant differences even within small geographical areas. Finally, differences in economic conditions will mean that optimal land management practices and responses to land degradation will vary even within similar ego-ecological areas.

National problems. The off-site effects of land degradation also vary substantially according to local conditions. The effect of sedimentation, for example, will depend on whether there is any valuable infrastructure downstream to be damaged. All watersheds do not feed the reservoirs of hydroelectric dams or irrigation systems.

Attention to site-specificity is closely connected to participatory approaches to land degradation control. Only the land users themselves are likely to have sufficient information on the agro-ecologic and socioeconomic conditions they face to be able to select the most appropriate management practice.

Only a small fraction makes its way into streams. And once in streams, sediment tends to be deposited on the streambed and mobilized several times before it ever reaches a reservoir or other vulnerable infrastructure. The time lag between erosion on a field and the consequent effect, therefore, is often measured in decades [Walling, 1988]. This has very important consequences for the evaluation of the benefits of interventions designed to reduce sedimentation.

Hydrological changes. Another major category of off-site effects which has caused concern are possible changes in hydrological patterns. If land degradation reduces infiltration rates, more of the rainfall will run off, resulting in:

- Reductions in dry season streamflow, reducing the availability of drinking water and reducing water supplies to irrigation.

- Increases in stormflow, resulting in flooding downstream and increased riverbank and streambed scour, thus increasing downstream sedimentation.
- Reductions in aquifer recharge.

Changes in the timing, volume, and velocity of water flow and groundwater recharge, can also alter natural lake, riverine, estuarine, and marine habitats, with adverse consequences for aquatic and riparian ecosystems. Most work on hydrological changes has focused on the consequences of land-use change in forested watersheds (see Chomitz and Kumari [1996] for a review). Information on the effects of land degradation on agricultural land on hydrological patterns is limited.

Water quality changes. In addition to changes in the quantity and timing of streamflows, land degradation can often have adverse consequences on water quality. Sediment increases the turbidity of streams, damaging fish and other aquatic lifeforms. It also increases the costs of potable water supply. In addition, nutrients lost from agricultural land can cause problems when they collect in waterways, by stimulating the growth of algae and other plants which deplete the available oxygen.

Land Degradation's Effects on Global Problems

Land degradation can contribute to global climate change, to loss of biodiversity, and to damaging international waters. Land degradation can affect problems of global concern in two ways.
- First, there can be *direct effects* of the land degradation processes themselves.
- Second, land degradation can cause *indirect effects* resulting from land users' responses to land degradation problems.

Climate Change

Emissions. Though not as important as non-agricultural activities, tropical agriculture is a significant source of greenhouse gases. Esti-

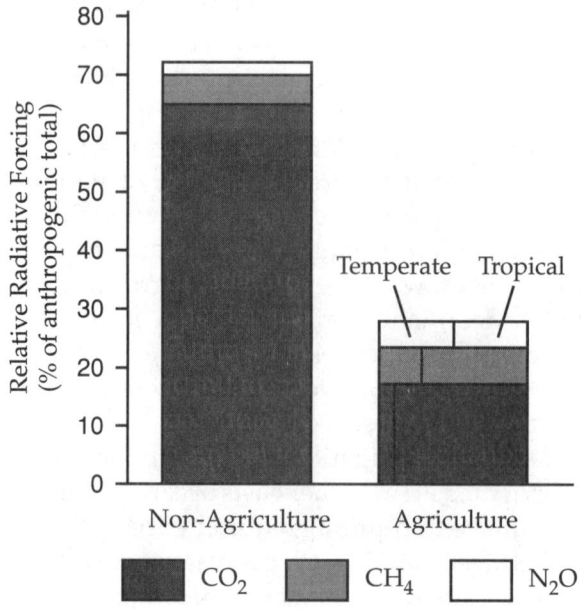

**Figure 3
Estimated relative contributions to global warming from agricultural and non-agricultural sectors**

Source: Adapted from Duxbury, 1995

mates of the impact of human activities on emissions of the main greenhouse gases—carbon dioxide (CO_2), methane (CH_4), and nitrous oxide (N_2O)—show a predominant influence of non-agricultural activities (Figure 3). These represent mainly the combustion of fossil fuels, which is accompanied by high carbon dioxide emissions. Within the agricultural sector, the main greenhouse gas emissions are of carbon dioxide and methane, and to a lesser extent nitrous oxide.
- Agricultural sources are estimated to account for about 30 percent of total carbon dioxide emissions; of this share, about 90 percent originates in tropical areas [Duxbury, 1995]. Deforestation is a major source of these emissions.
- Methane emissions are mainly from paddy rice production and livestock production.
- Nitrous oxide emissions originate mainly from fertilizer applications.

Terrestrial carbon sinks. Terrestrial ecosystems are an important carbon sink. While much

Figure 4
**Comparison of carbon stored in soil
and in above-ground biomass**

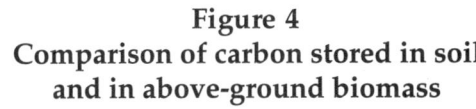

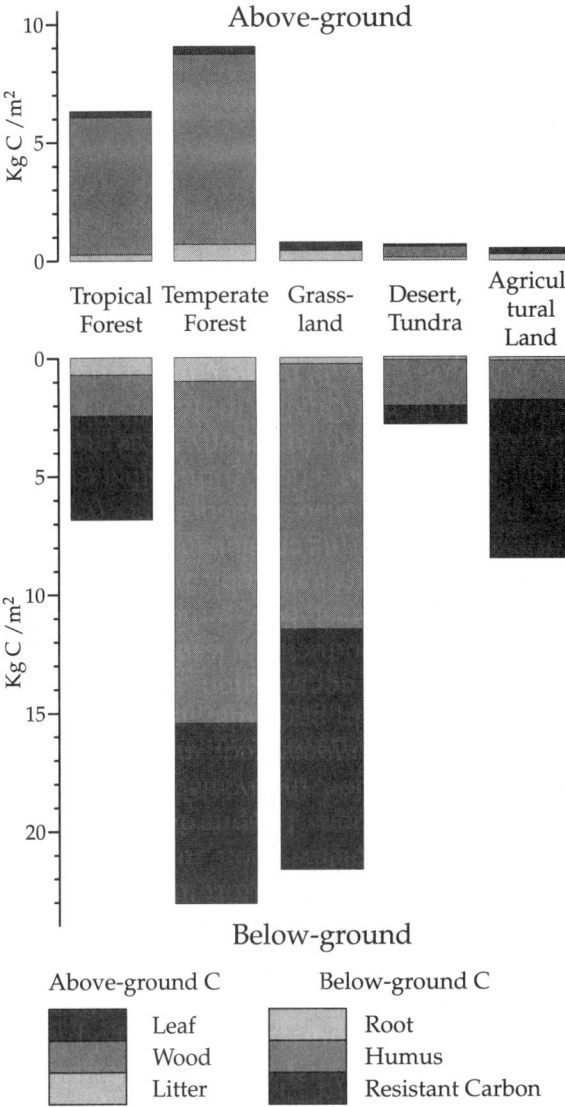

Source: Adapted from Goudriaan, 1993

have very high soil carbon, and indeed are thought to account for a substantial share of total carbon storage in the terrestrial pool. Soil carbon is lower on agricultural lands, but the amount stored is nevertheless much higher than might be expected from above-ground biomass alone. It should be stressed, however, that major uncertainties remain with regard to the role of the terrestrial carbon pool.

Africa's contribution to greenhouse gas emissions. Only 4 percent of Africa's arable land is used for rice cultivation, indicating that Africa's contribution to total methane emissions is likely to be minor, although this may be partly counterbalanced by the higher importance of pasture and crop residue burning practices and relatively low fodder digestibility on the African continent, all of which tend to increase methane emissions. African nitrous oxide emissions are also likely to be relatively low, given the low average level of fertilizer use (4.2 kg/ha versus 36 kg/ha in the tropics as a whole [Lal and others, 1995]). The remainder of this section, therefore, concentrates on carbon dioxide emissions.

Land degradation and climate change. Although agriculture as a whole is a significant contributor to global climate change, the question of interest here is the extent to which land degradation on agricultural land affects climate change.

- Does land degradation on agricultural land result in increased emissions of greenhouse gases?
- Does land degradation on agricultural land affect it's capacity to serve as a carbon sink?
- Can appropriate management enhance both land's productivity and its capacity to store carbon?

Unfortunately, data to answer these questions are scarce. Most attention has focused on emissions under specific land uses, and on the impact that land use change is likely to have (in particular, on the impact of converting forests to other uses). Very little work has been done on how land degradation within a given land use affects emissions.

attention has focused on carbon stored in above-ground biomass, storage in soils is substantial (see Figure 4). Sombroek and others [1993] estimate the stock of carbon stored in organic matter in the upper 1m of the world's soils to be about 1,220 Pg, or about 1.5 times the amount of carbon stored in biomass. Additional carbon is stored in deep soils as charcoal (50 Pg) and carbonate carbon (720 Pg). Grasslands can

Figure 5
The soil carbon cycle and some effects of degradation

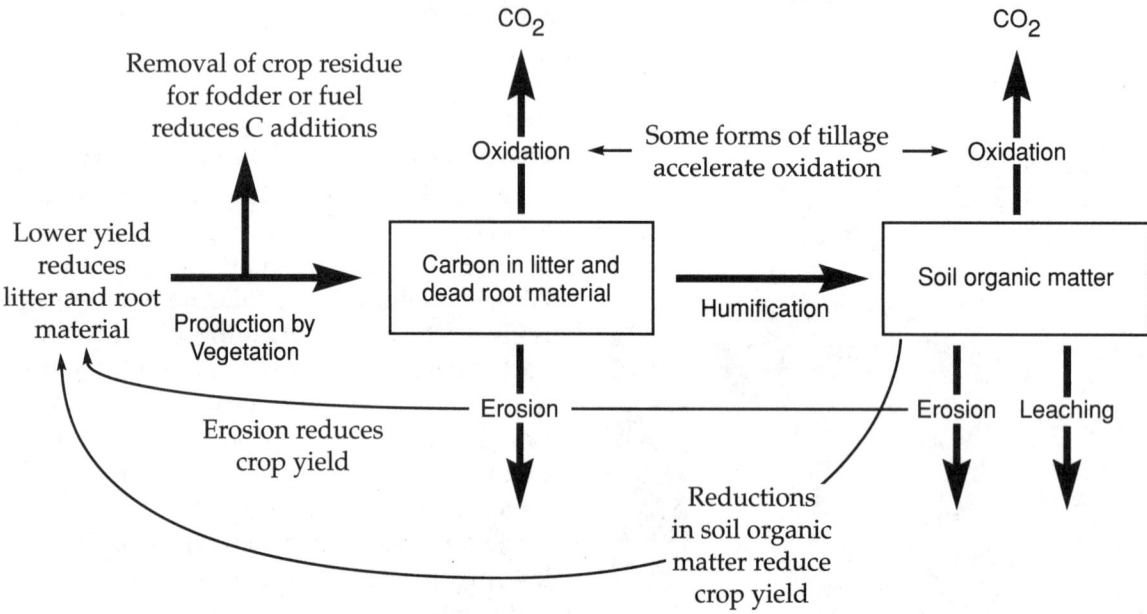

Carbon cycle in soils. Because the areas of interest in this paper (cropland and rangelands) tend to have relatively low above-ground biomass, the linkages between land degradation and climate change are likely to come primarily from changes in soil carbon. Figure 5 provides a simplified illustration of the carbon cycle in soils. The carbon balance in a soil at a given time is a function of the gains and losses of soil organic matter over time. Gains represent all forms of organic matter supply (litter, roots, crop residues, manure) while losses result from erosion, leaching, and decomposition of soil organic material. Decomposition is brought about by soil organisms using carbon compounds as a source of energy (oxidation), a process accompanied by the release of carbon dioxide. Decomposition rates depend on the characteristics of the organic material, of the soil organisms, and of the soil itself (for example, aeration, temperature, and moisture).

Effects of degradation. Soil organic matter plays a key role in soil fertility. The linkages between land degradation and carbon storage

in soils are complex, therefore, and can run in both directions. Some of the ways in which land degradation can affect, and may be affected by, the carbon cycle are illustrated in Figure 5.

• Some actions which *cause* land degradation can increase carbon emissions directly. For example, burning crop residues for fuel is thought to be an important contributor to fertility loss in many areas, since it prevents the return of many nutrients to the soil and reduces the build-up of soil organic matter. Some forms of tillage, particularly in arid and semi-arid environments, encourage oxidation of organic matter throughout the soil profile [Pieri, 1992], resulting in carbon being released to the atmosphere rather than building up soil organic matter.

• Some forms of degradation reduce soil carbon. Erosion carries away soil organic matter—often preferentially so since organic matter is highest in the upper soil layers which are most subject to erosion. Although this leads to a loss of soil carbon, however, it does not necessarily lead to increased emissions. Much of the carbon carried away by erosion may be deposited under conditions where it may be well preserved, such as in riverbeds and reservoirs [Duxbury, 1995; Tinker and others, 1996].

- The *consequences* of land degradation also affect the soil carbon cycle. Lower production of crops and pasture, whether it results from the damages of erosion, nutrient depletion, or other forms of degradation, will result in lower carbon inputs in subsequent periods (less root material, less leaf litter, less crop residue). This is an important linkage since it may contribute to a steady cycle of degradation.

The links between degradation and soil carbon are thus numerous and complex. It should be remembered, however, that reductions in soil carbon and increased emissions to the atmosphere are not necessarily synonymous. It should also be remembered that some forms of degradation can increase soil carbon. For example, pastures that are 'degraded' in the sense of being dominated by unpalatable grasses or woody species can accumulate substantial carbon stocks.

Estimating carbon losses. It does appear that land degradation is often correlated with increased carbon dioxide emissions and a reduced ability to store carbon. The magnitude of this effect is difficult to estimate, however. Although some information exists on the stocks of carbon under different land uses, there are few data on how these stocks *change* as a result of degradation. Some models exist that allow carbon balances to be calculated, but all have limitations. Complex process models such as the Erosion Productivity Impact Calculator (EPIC) [Williams and others, 1983] and CENTURY [Parton and others, 1992] usually require substantial data on soil and crop characteristics which are often not available in developing countries. Such models also usually require extensive calibration to local conditions before they can be used with confidence. Simpler models such as Soil Changes Under Agroforestry (SCUAF) [Young, 1989] are easier to use, but they

necessarily rely heavily on rough approximations. In particular, major uncertainties remain over both the rates of oxidation and the magnitude of the feedback between organic matter and yields.

Albedo. By reducing vegetative cover, land degradation can also change an area's albedo, providing an additional possible link to climate change. However, current research suggests that while such changes might have an impact on the local climate, they are unlikely to play a significant role in changing the global climate [Dickinson and others, 1996].

Biodiversity

Cropland. Croplands are substantially modified from their original, natural state, and their levels of biodiversity are generally substantially lower than those of natural habitats. Nevertheless, agricultural landscapes can contain considerable biodiversity [Pagiola and others, 1997]; indeed, biodiversity often plays a crucial role in agricultural production [Srivastava and others, 1996]. The concern here is whether land degradation might further reduce the remaining biodiversity.

Below-ground biodiversity. The main direct adverse effect of cropland degradation on biodiversity is likely to be on below-ground biodiversity. Diverse and abundant organisms help maintain soil fertility and productivity. This diversity is fundamental to soil quality —often called 'soil health'. Small organisms, such as insects and other invertebrates, play a vital role in developing and maintaining healthy soils, and help to maintain nutrient cycling, soil structure, moisture balance, and fertility. For example, *mycorrhizae*, which are fungi that live in symbiosis with plant roots, are essential for nutrient and water uptake by plants. Degradation of soil physical and chemical conditions can damage this biodiversity, about which relatively little is known.

The consequences of cropland degradation for biodiversity are not necessarily entirely negative. Abandonment of degraded lands may result in their ultimately reverting to their original condition, thus restoring to an extent the natural habitat. Whether this will occur depends partly on the availability of nearby habitat from which the abandoned areas can be recolonized, and partly on the condition of the abandoned fields.

Indirect effects. In most cases, the greatest impact of cropland degradation on biodiversity is likely to be indirect. By reducing productivity on existing agricultural land, degradation might force farmers to clear additional areas of natural habitat to maintain production. The fundamental arithmetic of agricultural production is that total production equals mean yield times area cultivated. Increasing demand for agricultural products, therefore, can only be met by increasing yields or expanding the area under cultivation. Houghton [1994] argues that much of the area being deforested is probably replacing land that is being abandoned after being degraded. This is a potentially serious problem, since conversion of natural habitat is the action that has the greatest impact on biodiversity [Pagiola and others, 1997]. It should be remembered, however, that land degradation is only one of many possible causes of agricultural expansion.

Rangeland. Rangelands often tend to be less modified from their natural state than cropland. They often contain a much greater proportion of their original biodiversity. Livestock often shares rangelands with considerable wildlife. The potential exists, therefore, for degradation to cause relatively more damage to biodiversity on rangelands than on cropland. There are many possible interactions between livestock and biodiversity in rangelands. Possible negative effects include the disruption of migration patterns; the introduction and propagation of diseases; competition for available food and water; and changes in forage species composition. Many of these adverse interactions are not degradation per se, however; biodiversity may be damaged even in a well-working, productive livestock system. Conversely, there is evidence that livestock can contribute to main-

---

Box 5. Biodiversity in drylands

When one thinks of biodiversity, the image that usually comes to mind is that of lush tropical rainforests. Biodiversity can also be quite high in drylands areas, however. Dryland ecosystems support a wide variety of fauna and flora, albeit at a relatively low density. Moreover, plants and animals in drylands are often characterized by considerable variation within species due to the need to adapt to the variability of the environment.

Since dryland plants and animals have adapted in ways that enable them to survive under harsh climatic conditions, they are a valuable source of genetic material to improve the tolerance of crops and livestock to drought and disease [Hassan and Dregne, 1997].

---

taining biodiversity [Mearns, 1996; de Haan and others, 1997].

Many of the aspects which are now thought to be particularly likely to cause pasture degradation [Steinfeld and others, 1997] are also likely to have an adverse impact on biodiversity. The encroachment of settled agriculture into the more favorable areas will limit their use by both livestock and wildlife. If access to areas that provide critical grazing or water at times of stress are restricted, for example, both will suffer. Agricultural encroachment and fencing that restricts the movement of livestock in response to seasonal climatic variations is also likely to impede the migration of wildlife. As both livestock and wildlife are restricted to smaller, often less favorable areas, competition between them is likely to be exacerbated.

International Waters

Many of the off-site consequences of land degradation that affect watercourses may be experienced beyond national borders. Sedimentation or flooding problems caused by degradation in an upstream watershed, for example, may affect a country downstream.

From the point of view of land degradation, damage to international waters are a special case of the off-site effects previously discussed. Technically, they are indistinguishable from the other off-site effects previously discussed. The main way in which damage to international waters differs from damage to waterways within the same country is that national policy-makers have no incentives to take them into account. The problems of identifying and measuring cause-and-effect relationships encountered in national off-site problems are likely to be even more severe in the case of international waters, since the limited data collection undertaken in the upstream and downstream countries may not be compatible.

Global Benefits of Land Degradation Control Activities

Land degradation problems can be addressed in many ways. A rough typology of land degradation control measures might include the following:

- Changing production technology: for example, by introducing practices such as minimum tillage or agroforestry.
- Adding conservation techniques to production systems: for example, by introducing terraces in cropland.
- Changing patterns of land use: for example, by relocating cropland and pastures to lower slopes and reforesting steeper slopes or by changing stocking rates on grazing land. Such efforts are often undertaken on a watershed scale.

If degradation is far advanced, such measures might have to be preceded by rehabilitation of the affected areas.

Just as land degradation on agricultural land can have a range of effects, measures to control it can also have a range of effects, at the field, national, and global levels. The specific range of benefits obtained will depend on the measure being used and on the conditions under which it is applied.

As discussed in the previous section, land degradation processes can have adverse global effects. A first benefit of land degradation control practices, therefore, is reducing or halting these adverse effects. To the extent that degra-

dation is reduced or halted, the problems discussed in the previous section will not occur. To the extent that land degradation is reversed, the global problems which had previously been generated will also be reversed. Increases in soil carbon, for example, are often a means of restoring productivity; they are also an *effect* of doing so. In either case, the soil's carbon sink function will no longer decline and may be enhanced. To the extent that land degradation is halted or reversed, any resulting pressure on natural habitats will also be alleviated.

In addition, some land degradation control practices can have additional benefits in and of themselves, by stimulating additional carbon sequestration and/or biodiversity over and above what might have occurred even in the absence of degradation. The following sections discuss some specific practices which seem particularly likely to have positive effects on problems of global concern.

Agroforestry

Reforestation could increase carbon sequestration substantially, but is unlikely to be feasible on a wide scale, since it would require forgoing crop production from the reforested areas. Agroforestry offers a compromise solution, since it allows increased carbon sequestration while continuing with crop production [Unruh and others, 1993]. Research on agroforestry practices has mostly been focused on the contribution they might make to the economic and ecological characteristics of farming systems. In addition to these benefits, agroforestry also has the potential of producing global benefits in terms of both climate change and biodiversity.

On-site and national benefits. Agroforestry practices can provide a range of on-site benefits. The trees can, of course, provide direct benefits in the form of products such as fruit, fuelwood, fodder, and timber. Combining trees with crop production can, in some cases, also improve crop productivity: shade can help reduce evaporation and provide a more favorable micro-

Table 1. Estimated average above-ground biomass and carbon content for different agroforestry types in Sub-Saharan Africa

Agroforestry type	Above-ground biomass (kg/ha)			Carbon content (kg/ha)		
	Low density	Medium density	High density	Low density	Medium density	High density
Silvopastoral						
0 - 400 mm rainfall	180	360	720	90	180	360
400 - 800 mm rainfall	175	350	700	88	175	350
> 800 mm rainfall	4,850	9,700	19,400	2,425	4,850	9,700
Fruit tree	3,000	7,500	15,000	1,500	3,750	7,500
Fuelwood	15,400	38,500	77,000	7,700	19,250	38,500
Shelterbelts		6,490		0	3,245	0
Timber trees	130,000	240,000	270,000	65,000	120,000	135,000

Source: Calculated from data in Unruh and others, 1993

climate for crops; trees can provide shelter from wind and reduce the erosive impact of rainfall; leaf litter can be used as mulch and improve soil quality by increasing soil organic matter. These are not, however, universal attributes: some trees compete with crops for nutrients and moisture, and some crops fare poorly under shade. From the farmers' perspective, the benefits of agroforestry derive from (i) additional production from the tree component; (ii) maintaining and/or improving the productivity of the crop component; (iii) diversification of production; and (iv) contribution to the overall farming system (for example, by providing fodder or income at a time when other sources do not) [Current and others, 1995]. Under some conditions, successful practices can also help abate off-site damages by reducing runoff and erosion.

Carbon sequestration. The tree component of an agroforestry system will sequester carbon from the atmosphere. The magnitude of this sequestration will depend on the specific tree species being planted, on their yield, and on the density of trees in the agroforestry system. Although there has been extensive research on agroforestry, little of it has measured the increase in carbon storage. Estimates of carbon sequestration in above-ground biomass can be obtained, however, from available information [Schroeder, 1993]. It is typically assumed that

the carbon content of biomass is 50 percent [Schroeder, 1993; Unruh and others, 1993; Young, 1989]. Table 1 provides estimates of average above-ground biomass and carbon content in a range of different types of agroforestry. Additional carbon would be stored below-ground, in the roots and in the soil. Little information exists on the magnitude of this storage, although some estimates suggest it might be substantial—perhaps as large as the above-ground storage [Unruh and others, 1993]. Given the low wood yields and the low planting densities possible under semi-arid and sub-humid conditions, the increments in carbon storage achievable in dryland areas would be at the lower end of the range shown in Table 1. The net gain in carbon storage would depend on the system that agroforestry replaces.

Biodiversity. Agroforestry systems tend to provide a more hospitable environment for biodiversity, both above- and below-ground. Some agroforestry systems can contain as much as half the species diversity found in neighboring primary forest [Thiollay, 1995]. Substantial biodiversity benefits are only likely, however, when agroforestry systems cover a relatively large area and are maintained for relatively long periods [Sanchez and others, 1996a].

Secondary effects. In addition to its direct effects on carbon storage on-site, agroforestry might have two secondary effects:

- To the extent that the tree component of the agroforestry system provides fuelwood, it would alleviate pressures on other fuelwood sources. In many cases, it will replace an unsustainable fuelwood source with a sustainable one. Since emissions from burning fuelwood are offset by new growth, the net flux to the atmosphere is zero [Unruh and others, 1993].

- If agroforestry succeeds in stabilizing and/or increasing yields on fields on which it is practiced, it will reduce pressures to clear additional areas for agricultural production [Schroeder, 1993]. This would prevent both the increase in emissions and the loss of biodiversity that would accompany such conversion.

Community-based Wildlife Management

Rangelands contain considerable biodiversity, most visibly in the form of wildlife. Land users have often been prevented from deriving any benefit from this wildlife by hunting or other means. In many countries, wildlife is legally considered property of the state. Moreover, wildlife is often felt to be competitive with productive uses such as grazing, by competing for forage or water or by propagating diseases. Not surprisingly, many land users actively seek to eliminate wildlife from areas used for grazing.

Recent years have seen many efforts to reduce the conflicts between human and wildlife use of the same land. These efforts can take two broad forms:
- Wildlife management as an alternative form of land use to grazing and/or cultivation; and
- Integration of wildlife and livestock in multi-species systems.

Both are related to land degradation, although in different ways.

Wildlife management. Many of the remaining areas with substantial wildlife tend to be in marginal areas unsuitable for intensive use [Kiss, 1990]. If used for cultivation or heavy grazing, they may degrade rapidly, damage

from which they will recover slowly if at all. Even a few years' use might be attractive to poor farmers or pastoralists, however, if the alternative is to derive no benefit whatsoever from a given area. Developing alternative ways for local communities to benefit from wildlands will thus both avoid degradation and preserve valuable biodiversity.

Various arrangements attempt to find ways for local communities to increase the benefits they receive from wildlife, so that preservation of wildlands becomes an attractive land use alternative to cultivation or grazing. The problems encountered tend to be primarily institutional and legal, rather than technical. In many cases, issues of property rights dominate. In others, new ways to derive benefits from wildlife need to be developed. Safari hunting or viewing are popular mechanisms for animal species, while royalties for bioprospecting offers some promise in the cause of plant species. Zimbabwe's CAMPFIRE program (Box 6) is perhaps the best-known example of an attempt to develop community-based wildlife management, but other examples can be found in many African countries.

Multi-species systems. Recent research indicates that livestock and wildlife are less competitive than was previously thought; recent research indicates that wildlife and livestock can be complementary to each other, and at times even symbiotic [Mearns, 1996]. The grazing 'overlap' between livestock and many wildlife species, for example, is now thought to be relatively limited [de Haan and others, 1997]. Moreover, as discussed above, it is now thought that many of the factors which tend to result in rangeland degradation also tend to adversely affect wildlife. This new understanding creates the potential for combining livestock and wildlife into multi-species systems.

Multi-species systems can make a fuller use of the ecological potential of spatially and temporally variable environments, and are often more suited to conditions in drylands than mono-species ranching [Cumming, 1994]. For such systems to be viable, however, problems such as encroachment of cultivation, fencing, and reduced access to waterpoints—which adversely affect both wildlife and livestock—must be addressed.

Summary

Land degradation on agricultural land is likely to affect problems of global concern in a number of ways, although the linkages are probably not usually as significant as those encountered in cases of deforestation. The main adverse global effects of land degradation on agricultural land are likely to be:

- *Climate change.* Land degradation on cropland and rangeland is likely to reduce the ability of soils to serve as a carbon sink and release carbon currently stored in soils to the atmosphere.
- *Biodiversity.* In rangelands, degradation may cause, or be related to through common causes, loss of plant and animal biodiversity. Degradation of both rangelands and cropland can also indirectly result in damage to biodiversity by increasing pressure to convert additional natural habitats to agricultural use.
- *International waters.* Land degradation can damage waterways in a number of ways, including sedimentation and changes in the quantity, quality, and timing of waterflow.

Just as land degradation can result in adverse global effects, land degradation control can result in positive global effects. In many

---

Box 6. CAMPFIRE: Community-based wildlife management in Zimbabwe

Zimbabwe's Community Areas Management Program for Indigenous Resources (CAMPFIRE) was established in 1989 in an effort to encourage land users to use wildlife sustainably by providing them with returns from doing so [Child, 1996]. Prior to the CAMPFIRE program, inhabitants of Zimbabwe's communal areas had no rights to exploit the often rich wildlife resources in their land, since wildlife was deemed the property of the state. Conversion of these fragile areas to agricultural use was often the only option open these communities, often resulting in land degradation. Local communities also had little incentive to resist poaching and, indeed, often participated in it themselves.

Under CAMPFIRE, communities can apply for the right to manage wildlife resources in their area. Management may involve both consumptive and non-consumptive uses, such as hunting (by the community or by tourists paying for the privilege) or photographic safaris. Revenues from such activities are channelled back to the community through their elected Rural Development Councils. In 1993, the program generated revenues of US$1.4 million. At present, the bulk of revenues originate from trophy hunting, particularly of elephants, although efforts are underway to develop tourism as an additional source of revenue.

---

cases, this positive effect takes the form of averted damages; in some, however, additional global benefits might be generated.

The following chapter turns to the issue of how attention to global problems might be better incorporated into land degradation control efforts.

3. Integrating Global Dimensions into Land Degradation Control Projects

The primary reason for efforts to control land degradation on agricultural land is to reduce, arrest, or reverse the field-level or national problems it is causing. Given the linkages between land degradation and problems of global concern detailed in the previous chapter, however, it is likely that global benefits will also be generated. This leads to two questions:

• How would land degradation control projects which incorporate global concerns differ from projects that do not?

• When land degradation control projects generate significant global benefits, how are costs to be shared between beneficiaries at different levels? In particular, since the GEF is the principal contributor to financing activities that generate global benefits, how are GEF's incremental costs guidelines to be applied to land degradation control projects?

This chapter examines the issues involved in integrating global dimensions into land degradation control projects. It begins by examining the motivation and incentives of individual land users. To be successful, any land degradation control program will need to obtain the cooperation of land users, and so an understanding of the constraints and incentives they face is necessary. Insufficient attention to this has meant that many land degradation control projects have failed to achieve their objectives. Previous efforts to address land degradation problems and the lessons which have been learned are reviewed in the following section. The chapter then discusses how attention to global problems can be integrated into land degradation control efforts—how they might

be modified to 'go the extra mile' to generate additional global benefits. Given the important role that the GEF plays in financing activities that bring global benefits, the chapter then discusses how land degradation control activities might fit in GEF's operational priorities and the application of GEF's incremental cost criteria. A number of pilot projects concepts which sought to integrate global dimensions in land degradation control activities are then reviewed.

Land Users' Motivation and Incentives

Production of crops and livestock depends on the decisions of individual producers. Farming requires constant, year-round attention by the producer. Moreover, the site-specificity of farming operations makes them heavily dependent on knowledge of the characteristics and idiosyncrasies of each location—knowledge which is typically only available to locally-based agents. Understanding the motivations of land users is critical, therefore, if patterns of resource use are to be understood and if appropriate policies to deal with any problems that might be identified are to be formulated.

That land users should adopt degrading practices or fail to adopt conservation practices that prevent degradation in the face of negative production effects has long puzzled and frustrated conservationists. But cultivation practices that damage the soil can also have beneficial aspects in terms of crop production, at least in the short term. Moreover, action to slow or arrest degradation through changes in crop and management practices or through adoption of

19

conservation techniques can be costly, both directly in terms of investment requirements or indirectly in terms of forgone production. Though sustainable practices may bring long-term benefits, they often have short-term costs. The critical question faced by land users is whether the long-term benefits of adopting sustainable practices make these costs worth bearing.

Land users generally have a wide range of possible responses to degradation. They might increase their use of fertilizer or other inputs, or change their land use practices to less damaging ones. A variety of reduced tillage practices exist, for example, which seek to reduce the damage to soil that conventional tillage can cause. Land users can also attempt to mitigate the effects of damaging practices by using off-setting conservation practices. Some of these conservation measures can themselves be productive; others might interfere with production, by reducing the effective area available for production or limiting the use of machinery.

Since land users will experience the effects of any on-site problems directly, they generally have a direct incentive to respond to them. And in many cases, they do. Almost all farmers in the Machakos and Kitui districts of Kenya, for example, have adopted some form of soil and water conservation practice [English and others, 1992; Tiffen and others, 1994; Pagiola, 1994]. Similarly, over half the fields farmed by small farmers in El Salvador have some form of conservation [Pagiola and Dixon, 1997]. When farmers do not respond to degradation problems, this might be the result of several reasons.

Costs and benefits of conservation. The available conservation options may simply not be cost-effective. For many years, research on conservation concentrated on physical structures such as terraces and drainage ditches. Such measures, while effective, are expensive both to build and to maintain. Moreover, since they often reduce the area available to cultivation, they incur an opportunity cost from forgone crop production in addition to their direct costs. In many cases, these costs are greater

than the benefits of averting degradation. Even cheaper conservation measures might not be worth undertaking, however, depending on their relative costs and benefits. Recent analyses of the economics of land degradation and conservation in many countries throughout the world have demonstrated that not adopting conservation measures can often be economically rational for farmers, even from a long-term perspective [Lutz and others, 1994; Pagiola, 1994; Pagiola and Bendaoud, 1995]. It is important to note that the relevant costs are those perceived by the land users, including the prices they actually pay for inputs and receive for outputs (which might be distorted by government policies), the opportunity cost of their own labor (frequently underestimated by analysts), and their own rate of time preference [Pagiola, 1993].

Policy-induced problems. Decisions to adopt conservation practices can often be heavily influenced by government policies, both by influencing the costs and benefits of conservation and by introducing other constraints:

- Until relatively recently, most developing countries had policies that discriminated heavily against agriculture. Resources were extracted from agriculture in a variety of ways: over-valued exchange rates, protection of competing sectors, price controls, and high direct taxation. A sample of 18 developing countries found that transfers out of agriculture during 1960-84 averaged 46 percent of agricultural GDP [Schiff and Valdés, 1992]. These policies made investments in agriculture, including conservation investments, less attractive.

- A variety of rules and regulations also affect behavior, although not always in the intended way. Under Mali's forest code, for example, any land left fallow for more than five years is considered part of the forest domain and hence government property; this clearly discourages land users from allowing productivity to regenerate naturally. Even rules intended to protect the environment can have adverse effects. Some countries have rules forbidding the cutting of trees [Current and others, 1995]. Although intended to prevent deforestation, such rules also discourage agroforestry and the planting of fuelwood lots. In many cases, rules and regulations often prove unenforceable, but can affect behavior by forcing farmers and others to avoid them (for example, by bribing enforcers to overlook infractions).

Tenure. When tenure security is uncertain, land users may be less likely to avoid activities that cause long-term damage or to undertake investments which bring long-term benefits because they are not sure that they will be able to enjoy these benefits. Here too, it is important to look at the problem from the perspective of individual land users. Absence of title, for example, does not necessarily imply insecurity of tenure. Few of the farmers in Kenya's Kitui district who had terraced their fields had titles to their land, for example [Pagiola, 1994].

Other problems. A variety of other problems can also constrain adoption of conservation measures. Undertaking conservation measures often requires investments, so lack of access to credit can prove to be a constraint. Poverty and the need to meet subsistence requirements is often thought to prevent poor farmers from undertaking conservation measures [Pagiola, 1995].

For a time, collectively-managed lands were thought to be peculiarly at risk of degradation, following Garrett Hardin's classic 'tragedy of the commons' argument. Research in recent years has shown, however, that communities can and do manage land resources successfully. When mismanagement of communal resources adversely affects the community itself, there are strong incentives to improve management. The problem Hardin was describing exists not so much on collectively-managed land but on *free access* land. Which is not to say that collectively-managed land is not more prone to problems, since coordination problems arise that are not present in the case of individually-managed resources. Moreover, traditional communal management regimes have often been undermined by social and technical changes and—sometimes deliberately—by government policies.

It should be noted that neither individual land users nor communities will take into account externalities in their decision-making. What is an externality from the perspective of an individual land user will differ, however, from the perspective of a community. Since communities cover a wider area than their individual members, some effects which are external from the perspective of the individual will be internal from the perspective of the community. To that extent, we might expect communities to respond to a broader range of environmental side-effects than individuals would.

In theory, national policymakers take both on-site effects of land degradation and off-site problems whose effects are felt nationally into consideration when formulating policies. (They may, however, value these problems differently than private agents.) They are unlikely to consider global effects, however, for two reasons. First, the bulk of the information they have available is likely to say nothing about any possible global impact of land degradation problems. Second, even when they understand the linkages between land degradation and global problems, national policymakers have no incentive to take global effects into account unless they are either committed to doing so by treaty obligations or compensated for doing so by the global community.

Approaches to Land Degradation Control

Diagnosis

Any land degradation control effort must begin with an assessment of the extent and severity of land degradation within a country and a diagnosis of its main causes. Much remains to be done in this area. Too often, perceptions of land degradation problems—both by the governments and by the World Bank—remain rooted primarily in anecdotal evidence rather than in a thorough review of available evidence. This is extremely worrying, since in several cases where a more careful analysis was attempted, the conventional wisdom was challenged substantially.

In-country. Countries affected by land degradation have undertaken a variety of assessments of the problems they face. Some particularly vulnerable countries have long-standing plans to combat land degradation. Mali and Niger, for example, both produced National Plans to Combat Desertification in 1985 and have since updated them. For most countries, however, the first attempt at a comprehensive assessment of environmental problems, including land degradation, was undertaken during preparation of their National Environmental Action Plans (NEAPs). NEAPs are intended to provide a framework to integrate environmental considerations into country economic and social development efforts. Signatories of the CCD affected by land degradation are to prepare a National Action Program (NAP), providing a new opportunity to review land degradation problems and prioritize interventions.

NEAPs often mention land degradation and —especially in Sub-Saharan Africa—often single it out as one of the most pressing environmental issues. As with other aspects of NEAPs, however, the analytical rigor tends to be low [Lampietti and Subramanian, 1995]. In many cases, they simply repeat the conventional wisdom regarding land degradation, with very little evidence to back it. The focus tends to be primarily on the easily-identifiable consequences of land-use change rather than on the more difficult to detect problems within land use systems. Very few NEAPs attempt to quantify the problems. Figures for rates of erosion or supposed productivity declines are sometimes cited, but upon examination either prove impossible to confirm or are found to be derived from single observations. Almost always, there is insufficient attention to the site-specificity of problems, with the analysis being carried out in terms of national averages. A common approach is to report the area that is

thought to be 'lost' to land degradation. These figures are generally uninformative. It is rare for an area to be degraded to the point that it is abandoned or otherwise 'lost'; it is much more common for an area to have its productivity slightly reduced by degradation. The impact of this degradation is less severe than on any 'lost' areas, but a far greater area is affected.

The analysis of the causes of land degradation is often particularly weak. That land users were ignorant of both land degradation problems and possible solutions was a common implicit—and often explicit— assumption.

World Bank. The World Bank has several instruments to aid its client countries in diagnosing land degradation problems, including Country Environmental Strategy Papers (CESPs), Agricultural Sector Reviews (ASRs), Economic and Sector Work (ESW) in the agricultural sector, and Environmental Impact Assessments (EIAs). A selection of these reports for African countries were reviewed for this study. Both the attention given to land degradation and the quality of analysis varied considerably. In many cases, discussion of land degradation has been relegated to an environmental section of sectoral work and CASs rather than being integrated into agriculture sector strategies. For example, the 1997 *Morocco Rural Sector Strategy*, although it reviewed other aspects of Morocco's rural sector in considerable detail, made almost no mention of land degradation problems, even though such problems are thought to jeopardize the future of irrigated areas because of reservoir sedimentation and to threaten productivity on rainfed lands—productivity which the *Strategy* hopes to increase substantially. Rather, a completely separate exercise was carried out later, in the context of an environmental review. The quality of analysis of land degradation problems also varied. At times, they do little more than repeat the conventional wisdom. Even where a more thorough analysis is attempted, it is often stymied by the lack of data. Yet in relatively few cases have follow-up efforts been made to fill the data gaps. Nevertheless, several examples of good practice can be found, including:

- *Ethiopia.* As part of the preparation of the CESP for Ethiopia, a comprehensive review of land degradation problems was undertaken. This review pointed to the need to revise previous perceptions of the nature and extent of land degradation [Bojö and Cassells, 1995].
- *El Salvador.* As part of a comprehensive review of prospects for and obstacles to growth in El Salvador, undertaken in cooperation with a Salvadoran NGO, a review of evidence of land degradation was undertaken and supplemented by analysis of a newly executed farm household survey. Here too, the results led to a substantial revision in the understanding of land degradation problems [Pagiola and Dixon, 1997].

Designing Solutions

Even where land degradation problems are thought to be the most important problem facing the country, attention to these problems tends to have a low priority within agriculture ministries. Land degradation, for all its supposed threats to agricultural productivity, is often relegated to environment ministries, which tend to be weaker institutionally and to have fewer resources and political clout. Moreover, many only have oversight responsibilities and cannot intervene directly.

Historically, the approach taken by most land degradation control projects has been to encourage land users to adopt some specific conservation measures. The proposed measures were generally selected centrally, on the advice of technical specialists. They were almost invariably assumed to be beneficial to land users, who were expected to readily adopt them once their benefits were demonstrated. Nevertheless, various forms of encouragement were often included, ranging from access to subsidized credit or inputs to outright subsidies of some or even all of the costs of implementation. At times, these efforts were further backed by legal obligations to either undertake certain measures or to not undertake others. It is symptomatic of this approach that project success was usually measured in terms of indicators such as 'linear meters of terrace constructed' and that almost no effort was made to measure either productivity or income effects.

It is fair to say that this approach has failed. In many cases, adoption of the recommended practices was low; in others, the practices were adopted temporarily but were soon abandoned once the project ceased. Seldom did adoption spread spontaneously to other land users who did not receive the subsidies.

- Many recommended conservation measures, while technically sound in terms of preserving land, were inappropriate to farmers' conditions. Many relied on the supposed existence of low- or zero-opportunity cost family labor for their implementation, for example, while others interfered with farmers' cultivation practices. Given the top-down, purely technical basis on which measures were selected, this is not surprising.
- Insufficient attention was paid to the policy environment.
- Insufficient attention was paid to the constraints land users faced and the causes of their use of degrading land use practices.

Indeed, far from achieving sustainable adoption of conservation measures, many projects seem to have created perverse incentives, leading to extensive construction of unnecessary conservation measures which were then rapidly abandoned once the project ended. In some cases, projects seem to have their incentive structure backwards. In El Salvador, for example, provision of subsidized credit was justified as allowing farmers to overcome the investment constraints associated with conservation investments. In practice, however, farmers appear to have been undertaking conservation measures so as to gain access to credit, making conservation a *cost* of obtaining credit rather than a benefit of doing so [Pagiola and Dixon, 1997].

Within the last decade, the emphasis of land degradation control projects has changed in two important ways.

- Much greater attention is now paid to changing the incentive framework within which land resource management decisions are made. This includes reform of the policy structure, with measures such as changes in price policy and liberalization of markets once dominated by government parastatals. It also includes efforts to improve the institutional framework, for example through tenure reform.
- Land degradation control efforts have been moving toward a much more participative approach, in which both the selection of solutions and their implementation are decided upon and executed in cooperation with beneficiary groups. The *Gestion des Terroirs* projects underway in several West African countries are a good example of this new approach (see Box 7).

Integrating Global Dimensions: Current Practice

Diagnosis. Given the low overall level of analysis of land degradation problems, it is not surprising that their relationship to problems of global concern has received even less attention. Biodiversity Strategy and Action Plans (BSAPs) are one vehicle through which consideration of the biodiversity aspects of land degradation problems might be achieved. BSAPs are a key instrument in countries' implementation of the Convention on Biological Diversity. One of their objectives is to identify threats to biodiversity and propose priorities for action. By and large, however, BSAPs have tended to focus primarily on biodiversity in relatively undisturbed areas, and have paid little attention to biodiversity in agricultural areas, degraded or otherwise [Pagiola and others, 1997]. Several mention agricultural encroachment as a pressure factor threatening biodiversity, but do not go beyond that.

Box 7. Gestion des Terroirs

The *Gestion des Terroirs* approach, also known as Community-Based Natural Resource Management (CBNRM), has been increasingly used in recent years, especially in Western Africa, where it was first developed. In this approach, communities design and implement a management plan for the area they regularly use (the *terroir*) with the help of a multidisciplinary team of technicians. The plan includes rules governing access to and exploitation of common resources such as pasture, forests, and water, and specific land improvement works, mainly on common lands but also on individual holdings.

The principles of the approach include:
- Management plans must be site-specific—there are no 'blueprints';
- Even where one activity, such as grazing, dominates the production system of a *terroir*, a multidisciplinary approach is needed;
- Management units must be based not only on social units (such as villages) but also on natural resources units (such as watersheds) that need to be treated together for management purposes;
- When a resource is shared by several communities, the management unit needs to include all the users.

Implementation of *Gestion des Terroirs* projects often requires substantial changes in government practices:
- Technicians and extension agents often have difficulty in internalizing the *terroir* concept and approach;
- Government departments tend to be poorly organized to deliver the required concerted and converging actions at the community level.

Although *Gestion des Terroirs* projects initially focused solely on natural resource management activities, they have gradually been broadened to include a wider range of activities, including social investments (which had been explicitly excluded in most early project designs). Recent projects typically provide for:
- Investments providing long-term economic and environmental benefits, such as soil erosion control, water-harvesting, soil fertility improvement, pasture improvement, natural forest management, sand dune fixation, tree planting, bush fire prevention, and windbreaks.
- Economic investments providing quick financial returns, such as wells, boreholes, and irrigation, construction of storage facilities for grain and production inputs, cereal banks, rural markets, animal health services, and processing units for food and animal feed.
- Social infrastructure investments, such as schools and health care centers.

Current projects. Relatively few current projects attempt to explicitly incorporate global dimensions. Of these, most take biodiversity as their entrypoint. Many protected areas projects, for example, have components that seek to improve productivity and/or reduce degradation in surrounding areas so as to alleviate encroachment into biodiversity-rich natural habitats. While such efforts will, if successful, undoubtedly have positive results on global problems, they are unlikely to play an important role in addressing a country's land degradation concerns since they only operate in very restricted areas. The Burkina Faso Environmental Management Project is an example of a project that addresses biodiversity protection from a land degradation entrypoint. In the southwestern provinces of Houet and Bougouriba, *terroir* management is combined with forest management. In these areas, forests are under severe pressure from massive land clearing by spontaneous settlers and fuelwood demand, threatening the remaining forest and wildlife. Integrated forest management and conservation plans for the protected forests were to be combined with *terroir* management plans for the adjacent community lands. The areas protected through the project include the Marc gazetted forest, the Hippo Pool forest (a UNESCO Biosphere Reserve), and the Nabéré gazetted forest and provisional wildlife reserve.

Sharing the Costs: The Role of the GEF

Calls have often been made for grant financing of land degradation control activities by international agencies, based on the presumed global benefits of such activities (for example, see Sanchez and others [1996b]). Since the Global Environment Facility (GEF) is the principal source of such financing, in practice the main issue is whether, and under what conditions, GEF grant financing might be obtained to assist in land degradation control projects. Both the Instrument Establishing the Restructured Global Environment Facility and the CCD assign the GEF an explicit role in this regard.

GEF financing criteria. GEF financing of land degradation control activities is subject to the following principles:
- Land degradation control activities are not eligible for GEF financing in and of themselves; they are only eligible for financing insofar as they relate to one or more of the GEF's focal areas (the relevant ones being climate change, biodiversity, and international waters). This is explicitly recognized in the CCD (art.20).
- The project should be consistent with the GEF's priorities, as described in its operational programs [GEF, 1997].
- As with other GEF projects, financing must conform to incremental cost criteria. GEF financing is not intended to finance activities which bring direct national benefits. This too is explicitly recognized in the CCD (art.20).
- The project must have a high probability of success and be a cost-effective means of generating the expected global benefits (as defined in the relevant GEF Operational Program or, in the case of climate change, Short-Term Measure).
- The project must be country-driven.

GEF's Priorities

Operational Programs. The GEF has developed ten Operational Programs (OPs) that describe the activities it is prepared to support in each of its focal areas [GEF, 1997]. Box 8 summarizes these Operational Programs and their relationship to land degradation control activities. It is important to realize that not all activities that generate global benefits are eligible for GEF financing. The GEF's Operational Programs set specific priorities and describe the activities the GEF supports in each focal area.

This point is particularly important in the case of measures to control climate change. Existing GEF Operational Programs on climate change focus on the introduction of new technologies and on the removal of obstacles to their adoption. As such, their relevance for land degradation control activities is limited. An operational program on carbon sequestration is under preparation. Until this program is adopted by the GEF Council, the GEF's Short-Term Measures are the main mechanism by which land degradation-related climate change mitigation measures might be eligible for GEF financing. The criteria for funding under this mechanism include a requirement that projects meet a cost-effectiveness criterion of sequestering carbon at a cost to the GEF of $10 per ton or less. This requirement is hard to meet in the case of land degradation in agricultural areas (as opposed to cases of deforestation) because of the lack of data on the magnitude of changes in carbon stocks caused by degradation.

Application of incremental cost principles

The application of incremental cost principles caused considerable confusion in the early years of GEF's work. As experience with their application has grown, the methodology for doing so has become clearer [GEF, 1996b; Hansen, 1997]. However, although efforts have been made to develop hypothetical examples of how incremental cost principles might apply to land degradation control projects [King and Kumari, 1997], specific experience with their application in this area is extremely limited. Problems in defining an appropriate baseline against which to measure incremental costs, already one of the most difficult issues in GEF projects, are likely to aggravated in land degradation control projects because of the presence of benefits at multiple levels and the scarcity of data.

An example can help to make the issues involved concrete. Consider a project to revegetate a degraded hillside. If left degraded, the hillside will generate few on-site benefits to land users, since its productivity will be low, and it may generate off-site damages to downstream areas. Revegetation would restore some of the area's productivity. It might make it once again usable for pasture, for example. It might also produce fuelwood and other valued products. Many of these benefits will accrue to

---

Box 8. GEF's Operational Programs

Biodiversity. Four of GEF's Operational Programs (OPs) are based on specific ecosystems:
- OP1 - Arid and Semi-Arid Zone Ecosystems
- OP2 - Coastal, Marine, and Freshwater Ecosystems
- OP3 - Forest Ecosystems
- OP4 - Mountain Ecosystems

The objectives of these OPs are the conservation and sustainable use of biodiversity and biological resources in the relevant ecosystem. Land degradation control efforts in these ecosystems are eligible for funding insofar as they address threats to the viability of the ecosystem, redress the damages of past degradation in biologically-sensitive areas, or develop sustainable use practices.

Climate change. The emphasis of the OPs on climate change activities (OP5, OP6, and OP7) is on the introduction of new technologies and the removal of barriers to their adoption, so they have limited relevance for land degradation problems. An OP on carbon sequestration is under preparation. Until such an OP has been adopted, the main eligibility criteria for GEF financing for land degradation activities that mitigate climate change are those of the short-term measures. Under these measures, among other criteria, projects must meet cost-effectiveness criteria. That is, they must mitigate a specified amount of greenhouse gas emissions for a given cost, typically a low unit abatement cost (approximately less than or equal to US$10 per ton of carbon).

International waters. Two other OPs with direct relevance to land degradation control activities are:
- OP8 - Waterbody-based Operational Program
- OP9 - Integrated Land and Water Multiple Focal Area Operational Program

OP8 focuses on seriously threatened waterbodies and on the most imminent threats to their ecosystems, while OP9 focuses more broadly on integrated management of land and water resources and on preventive rather than remedial measures.

---

local residents. Some benefits may also accrue to downstream residents, through a diminution of problems such as sedimentation or flooding. Revegetation might also generate some global benefits by sequestering carbon and providing a more hospitable habitat for biodiversity. Would such revegetation be eligible for GEF financing?

Figure 6
Application of GEF incremental cost principles

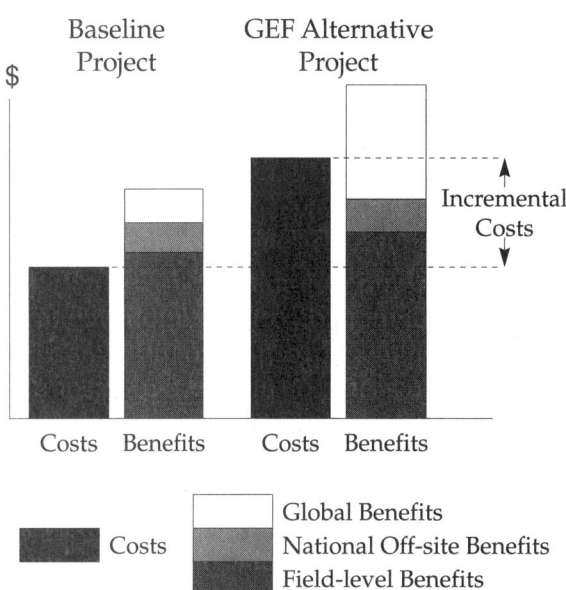

If revegetation is profitable based solely on the local and/or national off-site benefits that will be generated, it would not be eligible for GEF financing even if some global benefits are generated. Suppose, however, that the proposed revegetation project was modified to increase its contribution to alleviating global problems. For example, the hillside might be revegetated with indigenous species rather than with a monoculture of quick-growing exotics. This might result in greater biodiversity benefits by more closely recreating the original habitat. It might prove more costly than revegetating with a monoculture, however, both in terms of direct costs and perhaps also in terms of lower production of goods such as fuelwood. Alternatively, a larger area might be revegetated than originally envisaged. Some areas might have been left out of a revegetation project predicated solely on local and national benefits because they are unsuitable for grazing or too distant from population to make fuelwood production attractive. Yet benefits such as providing corridors between areas of natural habitat may make revegetating them attractive from

a biodiversity perspective. Such modifications of the original revegetation plan fall squarely within the incremental cost guidelines. As long as the global benefits being generated are sufficient and fall within the Operation Program guidelines, GEF financing would be appropriate. GEF financing would be available for the *incremental costs* of implementing the *GEF alternative* relative to the costs of implementing the *baseline project* (that is, the project the country would have undertaken based on national considerations alone). This is illustrated in Figure 6. If the GEF alternative project generates some additional national benefits compared to the baseline project, the incremental costs might be adjusted by subtracting the avoided costs to the country of generating these benefits.

In some cases, local and national benefits alone are insufficient by themselves to justify revegetation but the addition of global benefits would make the intervention profitable. The appropriate baseline for this project would be to do nothing. Under these circumstances, the same principles would apply: as long as the project is a cost-effective way of generating the expected global benefits in the context of an operational program, GEF would in general be willing to finance the incremental cost of the activity (once again, adjusted for the avoided costs of generating the local and national off-site benefits).

GEF financing is also available for activities which remove obstacles to adoption of practices which are expected to generate significant global benefits. For example, there might be a need for reforms to create a more enabling policy environment. Reform of the tenure system might be necessary to ensure that degradation problems do not recur following revegetation. Targeted research might be needed to either design appropriate revegetation schemes or to identify areas to be revegetated. Since these activities would also generate substantial local and national benefits, however, establishing an appropriate baseline is likely to prove difficult in such cases.

Implementation Issues

Land degradation control projects are likely to face difficult implementation problems as they attempt to address problems of global concern. In particular, two problems will pose themselves:

- How to design projects that take global concerns into consideration in a participative way; and
- How to secure the cooperation of land users in cases where the activities that generate global benefits are not in their direct interests.

As discussed previously, land degradation control projects no longer prescribe 'appropriate' conservation measures for land users to adopt. Rather, land users are offered a large choice of possible activities to undertake. In many instances, even the activities themselves are developed in cooperation with land users, through adaptive research. This approach is proving much more successful and sustainable than previous approaches, but it does pose problems for any efforts to integrate global concerns. Since the actual mix of activities that will be undertaken cannot be known in advance, identifying the likely global benefits will be difficult *ex ante*. Adapting the chosen activities so they generate additional global benefits (for example, by choosing species to be planted based on carbon-sequestration or biodiversity-enhancing criteria) will also be difficult.

From the perspective of individual land users, global benefits are as much an externality as national off-site effects. The same problems which have plagued efforts to undertake conservation measures to abate national off-site damages are likely to arise. Of course, if the conservation measures being envisaged are in the land users' own interest, then adoption and long-term maintenance will not be an issue. If this is the case, however, the measures may not meet GEF eligibility criteria. The approach taken by projects such as the *gestion des terroirs* projects in West Africa is to link certain conservation objectives to receipt of project-financed

investments in social infrastructure. For example, in the GEF-financed West Africa Pilot Community-Based Natural Resource and Wildlife Management Project in Burkina Faso and Côte d'Ivoire, progress on habitat and wildlife management is contractually linked to the provision of social infrastructure and other socioeconomic benefits from the project. Several of the project concepts discussed below have adopted this approach.

Pilot Project Concepts

In view of the limited experience in the preparation of projects that blend global environmental concerns with land degradation control, the World Bank and the International Fund for Agricultural Development (IFAD) have collaborated in developing a pipeline of projects in this nascent area of GEF operations. This effort, which was partly financed by a GEF Project Development Facility (PDF) Block B Grant of US$334,000, sought to clarify how global environmental benefits can be generated while addressing poverty and land degradation at the community level by developing a series of project concepts based on IFAD's pipeline of rural development projects. This section summarizes the initial lessons of this effort.

Selection Criteria. Table 2 summarizes the project concepts developed by IFAD. These project concepts were selected from within IFAD's pipeline of rural development projects using the following criteria:

- That the project directly address land degradation trends or address the physical production base of the rural poor in areas affected by or at risk of land degradation, such as arid, semi-arid, or dry sub-humid areas.
- That there appear to be links between land degradation processes in the project area and problems of global concern such as biodiversity protection.
- That potential GEF-financed activities be compatible with the proposed baseline project activities.

Table 2. Pilot project concepts integrating global dimensions in land degradation control

Land degradation problem	Baseline project	Global dimension	Possible response
Botswana: Community Based Natural Resources Management Project			
Rangeland degradation	Help local communities develop sustainable use practices	Degradation of one of the last large-scale migratory systems in the world	Emergency measures to rehabilitate the integrity of the southwestern wildlife system
Mali: Sahelian Areas Development Programme (SADeF)			
Unsustainable land use in the Interior Delta of the Niger River	Community-based natural resource management	Rich agro-biodiversity threatened by encroachment and unsustainable use	Targeted measures to conserve, rehabilitate, and promote the sustainable use of biodiversity
Jordan: National Program for Rangelands Rehabilitation and Development			
Rangeland degradation	Improved forage production and promotion of other viable productive activities	Damage to biodiversity; greenhouse gas emissions	Enhancing pasture regeneration to slow GHG emissions; measures to conserve arid zone biodiversity
Belize: Community-Initiated Agriculture and Resource Management Project			
Unsustainable land use and deforestation	Community-based natural resource management in agricultural areas	Encroachment into protected areas	Community-based natural resource management in non-agricultural areas
El Salvador: Rural Development in the North-Eastern Region			
Erosion and deforestation in hilly areas in the Bay of Fonseca watershed	Improved access to labour and credit markets; improved farming practices	Encroachment in remaining natural habitats	Reforestation of hillsides; Conservation measures in the buffer zones of areas with rich
India - Madya Pradesh: Madhya Pradesh Tribal Development Project			
Unsustainable land use and deforestation	Improved farming systems; small-scale irrigation; forestry activities; marketing of non-wood forest products	Encroachment into protected areas (11 national parks and 32 wildlife sanctuaries)	CBNRM and targeted activities to conserve unique plant and animal biodiversity
India - North East Region: Community Resource Management Project for Upland Areas			
Unsustainable land use and deforestation	Intensification of production in more favorable and accessible areas	Encroachment in natural habitats (area is designated a world ecological 'hot spot' due to its unique biodiversity)	Improvements to system of protected areas; creation of buffer zones; targeted research on natural regeneration
Mongolia: Arhangai Rural Poverty Alleviation Project			
Lake Baikal watershed is threatened by salinization due to poorly-managed schemes, overgrazing, and deforestation	Improved livestock and vegetable production; comprehensive rangeland monitoring	Lake Baikal and its surrounding area form a unique habitat for a range of aquatic and animal species	Afforestation; improved energy efficiency; development of a watershed management plan; strengthened management of protected areas

In addition, some projects were rejected for a variety of operational reasons, such as the stage of preparation making it impractical to add new components. It should be stressed that since project concepts were selected from IFAD's current pipeline of projects, they do not represent a systematic review of the various possible types of projects.

Botswana: Community Based Natural Resources Management Project

Wildlife in Southwestern Botswana has reached a critically low point, and its recovery is hindered by loss of critical wet season breeding range and dry season grazing range, impediments to migration, and high offtake. A number of water-dependent species have already been lost from the system. If current trends continue, large ungulates will be reduced to small populations in the protected areas. This would result in the loss of one of the last large scale migratory systems in the world, as well as of within-species genetic diversity. The proposed baseline project was to help local communities plan, prepare, and execute income-generating activities such as sale of hunting permits, wildlife and ecocultural tourism, and use and marketing of 'veld' products. The program also aimed to induce private sector involvement in tourism-related infrastructure and investment in the communities' capacity to manage wildlife resources and gazetting of protected areas.

The GEF component would have complemented the activities of the baseline by undertaking emergency measures aimed at reinstating essential components of the ecosystem, particularly through access to identified dry season range and water, so as to rehabilitate the integrity of the southwestern wildlife system. The benefits of these measures would in turn have been sustained by the CBNRM activities being promoted under the baseline program.

The proposed activities were consistent with the objectives of the GEF's operational program on Arid and Semi-Arid Zone Ecosystems (OP1), in that they provided for *in situ* conservation of biological resources threatened by pressure from more intensified use and drought, and encouraged sustainable use by helping to develop multiple use practices.

The program had strong links between baseline and GEF alternative activities. While the baseline focused on measures which would bring direct benefits to participating local communities and to the national economy, the GEF alternative would have undertaken complementary measures aimed at biodiversity conservation. Organizationally, the links between baseline and GEF alternative were strong since both were developed in parallel.

Although this concept was expected to have strong potential of obtaining GEF financing, Botswana's Ministry of Finance ultimately decided not to borrow for the baseline project.

Mali: Sahelian Areas Development Program

The Sahelian Areas Development Programme (SADeF) will follow an approach similar to the *Gestion des Terroirs* method (see Box 7 on page 24 above) by supporting local communities' efforts to plan, implement, and administer a variety of natural resource management initiatives. The Interior Delta of the Niger River is one of SADeF's areas of operation. This area provides the habitat for a variety of natural and human ecosystems, including a large diversity of wild and domesticated animal and plant genetic resources. Both the natural ecosystems and those modified by human influence have been degraded or are under severe pressure because of demographic growth, the progressive expansion of cultivated land, ever shorter fallow periods, soil salinization in irrigated areas, the increase of livestock herds, the reduction of pas-ture lands and pastoral corridors, and deforestation.

The baseline activities under SADeF will already result in global benefits by reducing the pressure on the remaining natural habitats of the Delta. However, preserving and restoring the region's rich and globally significant biodiversity, including animal and plant genetic resources, will require additional efforts and resources. The GEF alternative would complement SADeF activities with targeted measures to conserve, rehabilitate, and promote the sustainable use of the rich biodiversity in the Niger Delta. Specific activities would include:

- Integrating biodiversity conservation in SADeF's CBNRM activities.
- *In situ* conservation of domesticated and wild animal and plant genetic resources, by supporting local cultivators' production, distribution, and exchange of seeds of traditional landraces and by helping pastoral communities conserve traditional animal breeds through the establishment of Natural Traditional Breeds Centers.
- Targeted research on biodiversity resources, which will feed directly into the design and implementation of conservation activities, extension, and training.

The proposed activities are consistent with the objectives of the GEF's operational programs for Arid and Semi-arid Zone Ecosystems (OP1) and Coastal, Marine, and Freshwater Ecosystems (OP2), as well as on Integrated Land and Water Multiple Focal Area (OP9).

Preparation of this project is underway. A PDF B grant is being requested from the GEF to finance preparatory activities, including:

- Identification of gaps in the environmental information available.
- Preparation of the *in situ* conservation program, including prioritization of wild and domesticated species to be included, development of gender-specific incentive programs for use and preservation of indigenous knowledge that supports agrobiodiversity conservation, and definition of training requirements.
- Detailed design of the CBNRM Program, including development of a detailed 'menu' of options eligible for funding.
- Policy analysis of the underlying causes of land degradation.

Jordan: National Program for Rangeland Rehabilitation and Development

The project area is undergoing accelerating degradation because of high grazing pressure. Rangeland productivity is thought to have declined by up to 50 percent over the last 30 years, and is expected to decline further at a rate of 3 to 5 percent per year. The baseline project will address the poor socio-economic conditions of nomads by rehabilitating rangeland vegetation, with an emphasis on production of forage for livestock and promotion of other viable productive activities.

Several ways of supplementing the planned baseline activities with measures to enhance global benefits were examined, but none appeared promising:

- Enhancing pasture regeneration could have stimulated carbon fixation. However, the limited data available suggested that the project area was too small and regeneration too costly to be viable as an effective way of achieving global climate change benefits, and that it would be unlikely to meet the short-term response criteria of sequestering carbon for US$10 per ton or less.
- Efforts to re-establish the equilibrium between wildlife and domestic animals might have helped conserve biodiversity in this arid ecosystem. However, the global benefits were not well established. Although there is valuable biodiversity in the area, it was not clear to what extent land degradation trends threaten it. In addition, the synergy between the proposed biodiversity conservation efforts and the planned baseline activities was weak.
- Improving water harvesting practices and management could have helped decrease unsustainable use and aquifer depletion, but insufficient information was available to adequately evaluate the possible contribution of degradation control measures on international waters in the region.

The baseline project thus proceeded without a GEF component. Had the possibility of incorporating global aspects been raised earlier in project preparation, it might have been possible to fill some of the data gaps. This possibility will be re-examined when the next phase of the baseline project is prepared.

Belize: Community-Initiated Agriculture and Resource Management Project

Land degradation is a serious threat in Belize but is largely not yet a fact. However, the pressures on the natural environment are increasing rapidly: The rate of deforestation, for example, has accelerated considerably in the last few years and is expected to continue to do so. Pressures are particularly great in the south of the country. Existing agricultural practices only permit cultivation for short periods, and increasing population pressure is leading to shorter fallow periods.

The baseline project, to be implemented in Toledo and Stann Creek Districts in the south of the country, will address land degradation by developing sustainable production practices on land suitable for agricultural use, thus avoiding further declines in soil fertility or further expansion of the agricultural frontier into zones unsuited for agriculture. The project will also seek to improve farmers access to financial services and to technical and marketing services.

The GEF component would complement the baseline activities by approaching the same problem from the non-agricultural side: assisting communities in defining the lands that should not be used for agriculture, and developing ways to conserve them. GEF-financed activities would focus on the Sarstoon-Temash National Park, which includes the most pristine wetlands in the country. A CBNRM approach will be followed, with activities being drawn from a 'menu' of options based on Belize's biodiversity strategy. The problems at Sarstoon-Temash are very similar to those encountered in other protected areas in the countries, and the project is also intended as a pilot project which the approach proves successful, might later be replicated (the period of implementation of the GEF-funded activities is shorter than that of the baseline project to allow for such replication).

Because of the small-scale, targeted nature of the proposed activities, the GEF's Medium-Sized Grant program is being used. The proposed activities are consistent with the objectives of the GEF operational programs on Coastal, Marine, and Freshwater Ecosystems (OP2) and on Forest Ecosystems (OP3). The project area lies within the Meso-American Biological Corridor, a priority area for GEF biodiversity conservation activities.

El Salvador: Rural Development in the North-Eastern Region

The project area, in the Departments of La Union and Morazàn in the Bay of Fonseca zone, is the driest area in El Salvador. Much of the area's forests have already been lost, and the remainder are under pressure from fuelwood collection and clearing for agricultural use. Cropping areas are threatened by erosion due to steep slopes and lack of cover at the beginning of the rainy season, while pastures are subject to compaction and baring of soils due to overgrazing. The baseline project seeks to improve the living conditions of the rural poor, by improving their access to labor markets, supporting local financial institutions, and promoting non-degrading farming practices.

Possible activities considered for a GEF alternative included:

- Measures to protect biodiversity, such as rehabilitation of micro-watersheds and conservation measures in the buffer zones of areas with specific biodiversity needs, such as the Cacahuatique Mountains; and
- Testing and promoting more efficient cooking stoves as a means to prevent further deforestation caused by fuelwood use.

The strongest need for intervention concerns biodiversity conservation. There is clear evidence of biodiversity loss caused by loss of forest cover. However, there was only limited synergy between activities designed to protect the remaining biodiversity and the planned baseline activities, which mainly address areas already converted to cultivation. A targeted conservation program would probably provide a better vehicle to achieve biodiversity conservation objectives in this case. This is especially true since the baseline land degradation control project is already quite complex, with a disparate array of participating organizations. The case for activities designed to mitigate climate change is not as strong. Although deforestation results in release of carbon, most of the area's forests have already been lost. The link between land degradation in cultivated areas and climate change is not well established; nor is the extent to which promotion of energy-efficient cooking stoves would contribute to reductions in fuelwood consumption, or to reductions in greenhouse gas emissions. For these reasons, it was decided not to proceed with preparation of a GEF component at this time.

India - Madya Pradesh: Madhya Pradesh Tribal Development Project

Madhya Pradesh has the largest area of protected areas in India, including 11 national parks and 32 wildlife sanctuaries. Forests in the project area are being lost at a rate of 1.5-2.0 percent per annum. The main causes of the loss are increased demand for commercial and non-commercial forest products, including bamboo, and the expansion of cropping and livestock grazing. The baseline project had been designed to improve the quality of life of tribal groups and to prevent degradation of the environment by supporting a range of economic livelihood activities, including small scale irrigation, soil and water conservation, farming systems improvements, forestry activities, and support for marketing of minor non-wood forest products.

The GEF alternative would have aimed at strengthening biodiversity protection in the project area by investing in participatory institutional mechanisms for conservation and improved biodiversity management and by implementing targeted activities to conserve unique forest food and medicinal plant genetic resources and charismatic fauna in selected protected areas.

Development of a GEF alternative was hampered by uncertainty over the baseline project. Problems in the design of the baseline led it to be reformulated in 1997 and subsequently rejected in 1998, at which point preparation of the GEF alternative also ceased. Had work proceeded, it would have been necessary to better establish the global importance of the biodiversity that the project would have protected.

India - North East Region: Community Resource Management Project for Upland Areas

The project area, in the states of Meghalaya, Assam, and Manipur, in Northeastern India, has been designated as a world ecological 'hot spot' because of its unique biological diversity. Population growth has resulted in an increase in land area clearing for agricultural use and a shortening of the fallow period, eroding the sustainability of indigenous shifting cultivation system (*jhum*). The baseline project focuses on addressing the needs of the most vulnerable households through an integrated community-based approach for land use and productive support measures. These measures are designed to intensify production in the more favorable and more accessible areas. In addition, the project is supporting conservation activities for unique tribal cultural and religious areas of primary forest known as 'Sacred Groves'.

The GEF alternative would integrate biodiversity conservation with the baseline activities by enlarging existing protected areas, establishing new ones, and creating buffer zones to conserve the integrity of the parks. Targeted research and field testing of approaches for enhancing natural regeneration after *jhum* and to reduce the need for burning would also be undertaken. Efforts would also be made to integrate biodiversity considerations into areas of intensified agricultural production—for example by maintaining forest corridors between protected areas.

Even without the GEF component, the baseline project is likely to induce a reduction of pressure on existing protected areas, thanks to its agricultural intensification activities. In addition, the lifting of *jhum* pressure on land could result in longer fallow periods, during which significant biodiversity regeneration might take place. However, this would not address already existing damage, nor would it ensure adequate protection of natural habitats. By addressing these issues, the GEF alternative would substantially enhance the global benefits generated by the project. The proposed activities are consistent with GEF's Operational Program 4, on Mountain Ecosystems.

Although the proposed project and its GEF component appeared quite promising, both had to be put on hold because of a change in the national government, tensions between Central and North-East Region administrations, and civil strife in the region.

Mongolia: Arhangai Rural Poverty Alleviation Project

Lake Baikal is the world's deepest trough lake and holds one-fifth of the world's fresh water. The lake and its surrounding area form a unique habitat for a range of aquatic and animal species. Approximately two-thirds of the water flowing into Lake Baikal originates in Mongolia from the Selenge River watershed, which is threatened by salinization caused by poorly-maintained and inefficient irrigation schemes, overgrazing and trampling, and deforestation. The baseline project covers about one-third of the watershed and aims at raising the income of the poor herders through livestock development and vegetable production. It also introduces comprehensive rangeland monitoring and efforts to ensure that herds are kept to sustainable levels.

The GEF alternative would address the causes and effects of land degradation in the watershed by: (i) promoting energy efficiency in cooking and heating and afforestation; (ii) developing a water management plan for the watershed; and (iii) strengthening management of protected areas, so as to ensure quality and quantity of waterflows into Lake Baikal, thus helping to conserve its biodiversity.

Although the proposed GEF alternative appeared attractive from the perspective of protecting an internationally-important waterbody, it suffered from several problems. Most important, Lake Baikal is affected by extensive industrial pollution; land degradation only creates a small portion of the threats to the Lake's ecosystem. Although some efforts to address pollution are already underway, devoting substantial efforts to the more diffuse threats posed by land degradation is unlikely to be justified at this time.

Initial Lessons

Efforts to integrate global dimensions into land degradation control programs are still in their infancy. Nevertheless, some lessons are beginning to emerge.

Need to understand the underlying land degradation processes

The need for a clear understanding of the land degradation problems being confronted and of their causes was reinforced. One of the obstacles to the India Madhya Pradesh project was precisely that land degradation problems were not fully understood, leading to repeated redesigns of the baseline project, which in turn made it very difficult to design appropriate complementary activities to generate global benefits. In many of the projects reviewed for possible inclusion in the sample, the underlying land degradation problems were too ill-defined to allow consideration of the possible links to problems of global concern.

Establishing links to problems of global concern

Establishing whether there are links between land degradation and problems of global concern, and the exact nature and magnitude of those links, proved to be a significant obstacle in all cases. Development of all the project concepts was hampered by a paucity of relevant data.

Climate change. With the exception of cases in which deforestation was taking place, the prospects for activities designed to generate climate change benefits appeared to be very limited. In almost all cases, it appeared plausible that the proposed land degradation control activities would result in some reductions in carbon emissions, either by reducing losses from carbon stocks or by enhancing sequestration. In every case, however, the data were insufficient to establish the magnitude of this benefit and to demonstrate that it met the cost-effectiveness guidelines of the GEF's short-term climate change measures. The exception were cases involving deforestation, for which more data tended to be available. None of the project concepts examined involved agroforestry, so the extent to which activities which promote it might satisfy GEF climate change financing criteria remain unexplored.

Biodiversity. Linkages between land degradation and biodiversity proved much easier to establish in the sample of project concepts. These linkages generally took the form either of encroachment onto natural habitats adjacent to cultivated areas being degraded or of damage to wildlife that shares rangeland being degraded with livestock. There was relatively little information on possible losses of biodiversity within cultivated land. The SADeF program in Mali was the only exception; in this case, many of the diverse traditional landraces and cattle breeds found in the interior delta of the Niger River were found to be at risk from degradation processes. This is a reflection of the extraordinary agro-ecological diversity of the area, however, so it would be unrealistic to expect similar benefits in the majority of areas undergoing degradation.

Using land degradation as the entry point means that the biodiversity being conserved may not be the highest priority biodiversity. In cases such as the interior delta of the Niger River in Mali, the rangelands of Botswana, or southern Belize, land degradation problems coincided with areas of very rich biodiversity. In several other cases, however, the global importance of the biodiversity at risk was less obvious. In the India Madhya Pradesh project, for example, the degree of endemism among species in the project area was low even though there were several national parks in the area, and the area had not been identified as a priority in the national biodiversity strategy. This does not mean that efforts to protect this biodiversity would not be justified, but it does mean that a case for doing so must be prepared.

International waters. Only one of the project concepts had a significant international waters dimension: the Arhangai Project in Mongolia,

which addresses the problems of Lake Baikal. In this instance, however, it was judged that land degradation problems represented a relatively minor part of the waterway's problems, so an intervention on this basis was unlikely to be justified.

Integrating measures to address global problems with measures to address national problems

Even when the links between the global and the national and local dimensions of land degradation have been established, the question arises whether combining activities to combat them is desirable. The answer is not always yes. Two aspects need to be considered:

- The compatibility of the actual measures being proposed to combat the national and local problems and those required to address the problems of global concern.
- The compatibility of these activities from an organizational perspective.

Several projects showed considerable synergy between activities targeted at the local and national problems and those targeted at the global problems. Perhaps the best example of this synergy is that of the proposed Botswana CBNRM project. Sustainable use of rangelands is a precondition for implementation of the proposed GEF-financed wildlife conservation activities—without the planned baseline activities, the GEF-financed activities would be very unlikely to succeed, since the underlying problems would not have been addressed. In turn, implementation of the GEF-financed activities will bring benefits for the wildlife-based sustainable use activities. Likewise, in the Belize project, the proposed GEF-financed activities complemented those of the baseline by addressing the issue of appropriate management of community lands that fell within protected areas, while the baseline activities did the same for the surrounding agricultural areas. Conversely, in the El Salvador project the synergy was limited. The biodiversity loss problems being experienced were largely outside the agricultural areas targeted by the baseline project. Moreover, the activities required to protect this biodiversity had little in common with most measures contemplated under the baseline project, such as efforts to improve rural credit and labor markets.

The organizational synergy between measures to address national and global problems also varied. The synergy was high in the case of the Botswana project, since the same agencies were involved in both aspects of the problem. Conversely, the synergy was low in the El Salvador project, where the agencies that would have undertaken the biodiversity conservation activities were not the same as those involved in the baseline activities—indeed, adding more implementing agencies might actually have been harmful, since the project was already quite complex.

When either the substantive or the organizational synergy between measures to address the national and the global aspects of land degradation problems are low, it is likely that self-standing, targeted projects to address each problem separately would be more effective than attempting to integrate the two.

Implementation issues

Most project concepts addressed the problem of securing farmer participation by adopting variations of the now well-established CBNRM approach. The application of this approach to GEF-financed activities was most fully fleshed-out in the Mali project. Since CBNRM is a process rather than a blueprint, the specific activities will be defined during implementation, drawing from a 'menu' of options based on Mali's biodiversity strategy. Training for local communities will help ensure adequate local capacity to participate in developing natural resource management plans in and around protected areas and enter into 'contractual' agreements over the adequate conservation of areas with particular biodiversity value. Development of natural resources management plans and resulting project proposals and conservation measures would be financed out of a

CBNRM Fund with approval procedures similar to those for baseline activities. The key features of this plan involve the highly participatory nature of the choice of activities, and the use of explicit *quid pro quos* when conservation requires farmers to undertake activities which are not in their direct interests. How well these arrangements will work out in practice has, of course, yet to be determined.

One drawback of this approach at the time of project preparation is that it makes it difficult to establish *ex ante* the extent of global benefits that might be generated and their cost, since the exact activities that will be undertaken is not known. As long as the rationale for activities to generate global benefits was strong and their relationship to the baseline activities clear, however, these problems proved less severe than initially anticipated. The main difficulties seemed to lie in ensuring participation rather than in the nature of the activities to be implemented themselves. For the purposes of estimating incremental costs, representative activities can be used.

Determination of incremental costs

Incremental cost analyses were only carried out for some of the project concepts. As in other GEF projects, the distinction between costs that were eligible for financing and those that were not was the major point of discussion. In many cases, controversy arose primarily because of the mistaken expectation that any activity which generated global environmental benefits would be eligible. In several projects, including those in Botswana, Belize, Mali, and India's Northeast Region, important biodiversity conservation benefits are likely to be generated by the baseline activities, since they will tend to reduce pressures on natural habitats. Since these are incidental benefits of activities that are justified on the basis of the national benefits they generate, they are not eligible for GEF funding.

Once understanding was reached on this point, determination of incremental costs was relatively straightforward. Two factors, which may not always be present, eased the determination of incremental costs in the project concepts. First, since the project concepts were drawn from projects already under preparation, the baseline was already well defined. Second, most of the activities specifically designed to generate global benefits were *additional* to those foreseen in the baseline projects, making the incrementality obvious. If generating additional global benefits had involved *modifying* the baseline activities, apportioning costs would have been more complicated.

Summary

It should be stressed that these project concepts are only illustrative of the potentials and pitfalls of attempting to integrate attention to global benefits into land degradation control projects. Moreover, even the most advanced of the projects has only reached the advanced preparation stage; no doubt additional lessons will be learnt during implementation.

The projects which seem to lend themselves best to integrating global dimensions are those in which field activities are being carried out in specified areas. It is far more difficult to identify potential global dimensions of projects that seek to combat land degradation through policy reforms or support to research and extension. Because of the site-specificity of land degradation problems, it is very difficult to determine the nature and extent of links to global problems when only policy-level measures are contemplated. The synergy between measures to generate global benefits and the activities undertaken in these projects also tends to be low. Conversely, when field activities are being car-

ried out, identifying possible links to problems of global concern is substantially easier—particularly in the case of links to biodiversity.

The experience of the project concepts also suggests that establishing the nature and extent of linkages between land degradation on agricultural land and biodiversity is simpler than doing so for climate change. In the case of biodiversity, the main constraint is that the biodiversity at risk from land degradation on agricultural land may not have been as well studied as biodiversity in protected areas. In the case of climate change, the main constraint is that very few data exist on changes in emission resulting from changes within a given land use (as opposed to changes in emissions resulting from changes in land use itself. The data requirements are also more stringent in the case of climate change, due to the need to demonstrate cost-effectiveness.

The design of a GEF dimension for projects benefits from starting as early as possible in the project cycle—ideally at the identification stage. This will allow time for any data necessary for project development to be collected.

Conclusion

The land degradation question as a whole has not been marked by substantial success. More recent approaches, based on improved analyses of the causes and incentive structures within which land degradation occurs and employing a participative approach, appear to have a markedly improved chance of success. However, integrating global dimensions into such efforts will not be an easy task.

4. Conclusions and Next Steps

There are clearly important direct and indirect linkages, in many instances, between land degradation on agricultural land and problems of global concern. Although these linkages are not likely to be as strong as those encountered in cases of deforestation, land degradation on agricultural land can:

- Reduce the ability of soils to serve as a carbon sink and release carbon currently stored in soils to the atmosphere.
- Lead to loss of plant and animal biodiversity in rangelands;
- result in increased pressure to convert natural habitats to agricultural use.
- Damage international waterways.

The strength of these linkages will vary substantially from case to case, however. In cases where land degradation results in adverse global effects, land degradation control can result in positive global effects.

These linkages, when they are present, open the possibility that efforts to control and reverse land degradation and efforts to mitigate global problems can be mutually supportive. Equally clearly, however, explicitly incorporating global considerations into land degradation control efforts will not be easy.

The two main difficulties likely to be encountered by efforts to incorporate global considerations into land degradation control activities are:

- *Information.* If land degradation processes are to be better addressed, there is a clear need for an improved understanding of their extent, nature, and severity. In terms of linkages to global problems, although sufficient information is available to indicate that land degradation can, under many conditions, lead to adverse global effects, specific information on the nature and magnitude of these effects is extremely scarce.
- *Implementation.* Land degradation control programs have proven extremely difficult to implement unless on-site benefits are high. When local and global benefits are highly correlated, this does not pose a problem for the achievement of global benefits, although it does imply that there may be little need for outside financing (either from national or global sources). When the correlation is low, so that different or additional interventions are required to secure global benefits, however, the problems might be substantial. Just as it has proven extremely difficult to undertake land degradation control efforts that mitigate national off-site damages in a sustainable way, it will be difficult to mitigate global off-site damages in a sustainable way.

Given these difficulties, the next steps in the development of effective programs to address land degradation and global concerns need to be undertaken on two parallel tracks. First, efforts need to be made to plug some the gaps in the data—on both local and national problems and on any resulting global effects. Second, there needs to be experimentation with different approaches to incorporating global concerns in land degradation control projects.

Research Needs

Any effort to develop projects that address both land degradation and global concerns must be based on a clear understanding of the underlying land degradation problems, their causes, and their effects at both the farm and national levels, as well as on an understanding of their linkages to problems of global concern. Yet accurate and reliable information on both of these aspects is often not available.

National research systems. Given the site-specificity of land degradation problems, National Agricultural Research Systems (NARS) clearly have an important role to play in understanding its effects on problems of both national and global concern. However, NARS have traditionally been oriented almost solely towards short-term productivity considerations, an approach that many maintain to this day. In addition, many NARS are institutionally weak and short of resources. Their links to extension have also often been weak.

International agricultural research. The whole focus of the international agricultural research centers in the CGIAR system was also traditionally solely on productivity enhancement; environmental and natural resources concerns were not addressed and farmers' participation was not sought. Although productivity growth is still the main objective, environmental concerns are now also being considered, for example through farming systems research. In 1996, the CGIAR centers spent about US$49 million (about 17 percent of overall expenditures) on soil and water conservation research. About 35-40 percent of soil and water conservation research is being carried out in high potential areas (irrigated areas and rainfed lowlands). Funding for soil conservation research in the CGIAR system is 4 times that for water conservation. About 70 percent of CGIAR research focuses on on-site effects at the field or farm levels; about 30 percent looks at effects at the community and watershed levels.

In an effort to improve the availability and quality of data on land degradation, the World Bank is collaborating with several other organizations in the Land Quality Indicators program (Box 8).

In addition to the research required to develop appropriate responses to land degradation at the local and national level, research

---

Box 9. Land Quality Indicators

Land quality is the condition or health of the land relative to its capacity for sustainable land use and environmental management. At present, however, few indicators are available that allow changes in land resources to be monitored and evaluated. The Land Quality Indicators (LQI) Program being developed by a coalition of international agencies including the World Bank, FAO, UNDP, and UNEP, aims to help fill this gap [Pieri and others, 1995].

LQIs are needed to address major land-related issues such as land use pressures, land degradation, and soil and water conservation. The LQI program is developing a core set of indicators to achieve this. Once developed, LQIs will be used for policy and program formulation for district, national, and global assessment, environmental impact monitoring, and to promote technologies, policies and programs to ensure better use of natural resources and sustainable land management.

The LQIs being developed for immediate application include:

- *Nutrient balance*: describes nutrient stocks and flows under specific land uses.
- *Yield gap*: compares current with potential yields.
- *Land use intensity*: describes the impact of intensification on land quality.
- *Land cover*: describes the extent, duration, and timing of vegetative cover.

Indicator guidelines for nutrient balance and yield gap have been prepared; guidelines for the other indicators are in various stages of development. Additional LQIs that attempt to capture soil quality, land degradation, and agrobiodiversity are still in the research stage. Work is also underway on developing indicators of the extent of carbon storage (above- and below-ground) under different land uses.

---

is also required on the global dimensions of land degradation. This is likely to prove more difficult to organize on a systematic basis, since national research organizations have no direct incentive to undertake such research and, indeed, often have insufficient resources even for their national research objectives. Mindful of this, the GEF has already financed targeted research in areas where the potential contribution to global problems is significant. For example, the Alternatives to Slash and Burn program examines the greenhouse gas emissions and biodiversity effects of slash-and-burn practices and alternative land use systems in tropical forests [Tinker and others, 1996].

Global dimensions in World Bank Agriculture Sector Work. As noted earlier, attention to land degradation problems has not always been fully integrated into the Bank's agriculture sector work. In countries threatened by degradation, greater efforts need to be made to understand the nature and severity of degradation problems and the threats they might pose to agricultural sustainability and to integrate efforts to address them into agriculture sector strategies.

The need to address global dimensions, where relevant, adds an additional layer. Through the Global Overlay Program, the World Bank is working to ensure that attention to global problems becomes a regular part of mainstream Bank work. Guidelines for Climate Change Overlays have already been prepared [World Bank, 1997], and best practice guidelines for biodiversity overlays are under preparation [Pagiola and others, 1997]. The Bank needs to strengthen its agriculture sector work to effectively help developing country partners mainstream climate change mitigation and biodiversity conservation in planning for this sector.

In a local context, Bank staff interacting with country sector planners need to be able to address several questions:

- What are the extent, severity, and causes of the land degradation problems facing the country?
- What are the effects of degradation at the farm and national levels?
- How do the specific land degradation processes at work affect problems of global concern?

Of course, full and complete answers to these questions will not be possible in most cases, given the current state of knowledge. Even preliminary efforts to answer them can often provide considerable information and help target research efforts towards the most important gaps. It is also important to bear in mind that since land degradation problems, and their global dimensions, tend to be highly site-specific, it is not usually necessary to have complete information on conditions in the entire country. What is needed is information on the specific area where interventions are being envisaged.

Operationalizing the Global Dimensions of Land Degradation

The World Bank has already devoted considerable resources to assist client countries undertake land degradation control activities [World Bank, 1996, 1998]. During 1990-95, the Bank financed 108 projects (for a total of US$13.4 billion) that aimed at improving land management in dryland areas; of these, 34 projects (for a total of US$2.2 billion) dealt primarily with land degradation. Additional projects are in the pipeline.

Soil fertility initiative. In the Africa region, where land degradation problems are most severe, a Soil Fertility Initiative (SFI) has been launched by the World Bank, in partnership with a number of international, public, and private organizations. The objectives of the initiative are to help reverse the cycle of soil degradation in Sub-Saharan Africa and increase the sustainability of agricultural production. Soil fertility improvement action plans will be developed and incorporated in country assistance strategies, and soil fertility components will be included in relevant Bank-financed operations. This work program began with several pilot countries, in which the activities are being tested and refined. These include Guinea, Madagascar, Malawi, Mali, and Uganda. The FAO is participating in the design and supervision of specific soil fertility project components.

It is quite likely that many of the land degradation control projects already planned or underway will generate global benefits. Efforts will be made to also incorporate explicit attention to global problems in forthcoming land degradation projects, with financial support from the GEF when appropriate.

The point of departure of any effort to incorporate global dimensions into land degradation control efforts, must be a well-thought out strategy to address the local and national aspects of the problem. This requires a clear understanding of the nature, extent, and severity of land degradation problems, their causes, and their effects at both the farm and national levels. It also requires, as discussed previously, a clear understanding of the incentives and constraints faced by land users.

At this stage, it should be possible to determine whether the measures already envisaged will be sufficient to address most global problems originating at the site, or whether additional or different measures would be required to do so. Here too, project designers need to ensure that the measures being contemplated are consistent with land users' incentives and constraints. Since such measures will usually involve what is, from the perspective of land users, the abatement of externalities, some mechanism will often be required to compensate them for undertaking the proposed measures. As discussed in Chapter 3, this will not be an easy task.

In cases where an application for GEF financing for the incremental cost of activities designed to address problems of global concern is being envisaged, it will then also be necessary to determine

- How the proposed measures are likely generate benefits in the GEF's focal areas (climate change, biodiversity conservation, and protection of international waters);
- Whether the proposed activities are consistent with GEF's Operational Programs; and
- What the expected incremental costs from the proposed activities are.

The project concepts developed by IFAD in collaboration with the World Bank provide some initial examples of what is possible in this field.

Conclusion

For many countries—and in particular for many African countries—land degradation on agricultural land is posing substantial threats to sustainability, economic growth, and the welfare of the rural population. Strong efforts to combat land degradation are justified on these grounds alone. In some cases, reduction of problems of global concern such as mitigation of climate change or conservation of biodiversity provide an additional reason to combat degradation. At times this may require additional or different measures than if local and national considerations were the only ones involved.

In cases where there are strong linkages between land degradation and problems of global concern, efforts to combat both can be mutually supportive. It is important, however, to avoid having the tail wag the dog. The primary motivation for land degradation control efforts will remain the local and national benefits that can be derived thereby. Linkages to global problems are not always present, or may not be sufficiently strong to warrant specific attention.

Glossary

Agroforestry - A land use in which trees or other woody perennials are incorporated into fields used for crop or animal production.

Biodiversity - Short for biological diversity; it encompasses the variability among living organisms from all sources, including, among others, terrestrial, marine and other aquatic ecosystems and the ecological complexes of which they are part; this includes diversity within species, between species, and of ecosystems.

Extensification - Increasing agricultural production by expanding the area under cultivation.

Gestion des Terroirs - Approach to community-based land resource management developed in West Africa. Under this approach, communities design and implement, with the assistance of a multidisciplinary team of technicians, a management plan for the territory on which they live permanently or seasonally and on which they regularly carry out agricultural and/or livestock production activities and/or other renewable resources activities (the "terroir").

Greenhouse Gas - Chlorofluorocarbons (CFCs), carbon dioxide (CO_2), methane (CH_4), and nitrous oxide (N_2O) are the main greenhouse gases which are significantly increased by human activity.

Intensification - Increasing the use of inputs and/or changing land use so as to increase productivity (output per unit of land).

Natural habitat - Land and water areas where (i) the ecosystems' biological communities are formed largely by native plants and animal species, and (ii) human activity has not essentially modified the area's primary ecological functions.

Off-site effects - Effects of a land use change that are felt outside the area on which the land use change is carried out.

On-site effects - Effects of a land use change that are felt within the specific area on which the land use change is carried out.

References

Agro-Concept. 1995. *Plan National d'Aménagement des Bassins Versants*. Rabat: Ministère de l'Agriculture et de la Mise en Valeur Agricole.

Behnke, R.H. Jr, and I. Scoones. 1993. "Rethinking Range Ecology: Implications for Rangeland Management in Africa." In R.H. Behnke Jr, I. Scoones, and C. Kerven (eds), *Range Ecology at Disequilibrium*. London: Overseas Development Institute.

Bojö, J., and D. Cassells. 1995. "Land Degradation and Rehabilitation in Ethiopia: A Reassessment." AFTES Working Paper No.17. Washington: World Bank.

Brady, N.C. 1986. *The Nature and Properties of Soils*. Ninth Edition. New York: MacMillan.

Brown, L.R., and E.C. Wolf. 1984. *Soil Erosion: Quiet Crisis in the World Economy*. Worldwatch Paper No.60. Washington: Worldwatch Institute.

Chile, B. 1996. "Zimbabwe." In E. Lutz and J. Caldecott (eds), *Decentralization and Biodiversity Conservation*. Washington: World Bank.

Chomitz, K., and D.A. Gray. 1996. "Roads, Land Use, and Deforestation: A Spatial Model Applied to Belize." *World Bank Economic Review*, Vol.10 No.3, pp.487-512

Chomitz, K., and K. Kumari. 1996. "The Domestic Benefits of Tropical Forests: A Critical Review." Working Paper GEC96–19. Norwich: CSERGE.

Cleaver, K.M., and G.A. Schreiber. 1994. *Reversing the Spiral: The Population, Agriculture, and Environment Nexus in Sub-Saharan Africa*. Washington: World Bank.

Clark, E.H., H.J. Haverkamp, and W. Chapman. 1985. *Eroding Soils: The Off-Farm Impacts*. Washington: Conservation Foundation.

Crosson, P.R., and A.T. Stout. 1983. *Productivity Effects of Cropland Erosion in the United States*. Washington: Resources for the Future.

Crutzen, P.J., and M.O. Andrea. 1990. "Biomass Burning in the Tropicas: Impact on Atmospheric Chemistry and Biogeochemical Cycles." *Science*, No.250, pp.1669-1678.

Cumming, D.H.M. 1994. "Are Multispecies Systems a Viable Landuse Option for Southern African Savannas?" Paper presented at the International Symposium on Wild and Domestic Ruminants in Extensive Land Use Systems, Berlin, 3-4 October 1994.

Current, D., E. Lutz, and S.J. Scherr. 1995. "The Costs and Benefits of Agroforestry to Farmers." *World Bank Research Observer*, Vol.10 No.2, pp.151-180.

de Haan, C. H. Steinfeld, and H. Blackburn. 1997. *Livestock-Environment Interactions: Finding a Balance*. Brussels: European Commission Directorate-General for Development.

Dickinson, R.E., V. Meleshko, D. Randall, E. Sarachik, P. Silva-Dias, and A. Slingo. 1996. "Climate Processes." In Intergovernmental Panel on Climate Change, *Climate Change 1995: The Science of Climate Change*. Cambridge: Cambridge University Press.

Dregne, H.E., M. Kassas, and B. Rozanov. 1992. "Status of Desertification and Implementation of the United Nations Plan to Combat Desertification." *Desertification Control Bulletin*, Vol.20, pp.6ff.

Duxbury, I. 1995. "The Significance of Greenhouse Gas Emissions from Soils of Tropical Agroecosystems." In R. Lal, J. Kimble, E. Levine, and B.A. Stewart, eds, *Soil Management and Greenhouse Effect: Advances in Soil Science*. Boca Raton: CRC Press.

English, J., M. Mortimore, and M. Tiffen. 1994. "Land Resource Management in Machakos District, Kenya, 1930-1990." Environment Paper No.5. Washington: World Bank.

Global Environment Facility (GEF). 1996a. "A Framework of GEF Activities Concerning Land Degradation." Washington: GEF.

Global Environment Facility (GEF). 1996b. "Incremental Costs." GEF Council Report GEF/C.7/Inf.5. Washington: GEF.

Global Environment Facility (GEF). 1997. "GEF Operational Programs." Washington: GEF.

Goudriaan, J. 1992. "Biosphere Structure, Carbon Sequestering Potential and the Atmospheric ^{14}C Carbon Record." *Journal of Experimental Botany*, Vol.43, pp.1111-1119.

Hansen, S. 1997. "Incremental Costs as it Relates to Land Degradation: Current Practices and Methodological Issues." Paper presented at the STAP Meeting on Incremental Costs and Global Benefits in Land Degradation, Amsterdam, 17 June 1997.

Hassan, H., and H.E. Dregne. 1997. "Natural Habitats and Ecosystems Management in Drylands: An Overview." Environment Department Paper No.51. Washington: World Bank.

Houghton, R.A. 1994. "The Worldwide Extent of Land-use Change." *BioScience*, Vol.44 No.5, May, Special Issue on the Global Impact of Land-Cover Change, pp.305-313.

Kellenberg, J., and D. Cassells. forthcoming. "Incorporating Global Externalities in Forest Conservation and Management." Global Overlays Program. Washington: World Bank.

King, K. and K. Kumari. 1997. "Incremental Costs of Focal Area Activities that also Address Land Degradation." Paper presented at the STAP Meeting on Incremental Costs and Global Benefits in Land Degradation, Amsterdam, 17 June 1997.

Kiss, A. 1990. "Living with Wildlife: Wildlife Resource Management with Local Participation in Africa." Technical Paper No.130. Washington: World Bank.

Lal, R.C. 1987. "Effects of Erosion on Soil Productivity." *CRC Critical Reviews in Plant Sciences*, Vol.5 No.4, pp.303-367.

Lal, R., and T.J. Logan. 1995. "Agricultural Activities and Greenhouse Gas Emissions from Soils in the Tropics." In R. Lal, J. Kimble, E. Levine, and B.A. Stewart, eds, *Soil Management and Greenhouse Effect: Advances in Soil Science.* Boca Raton: CRC Press.

Lampietti, J.A. and U. Subramanian. 1995. "Taking Stock of National Environmental Strategies." Environment Department Paper No.10. Washington: World Bank.

Lutz, E., S. Pagiola, and C. Reiche. 1994. "Cost-Benefit Analysis of Soil Conservation: The Farmers' Viewpoint." *The World Bank Research Observer*, Vol.9 No.2, July, pp.273-295.

Magrath, W.B., and P. Arens. 1989. "The Costs of Soil Erosion on Java: A Natural Resource Accounting Approach." Environment Department Working Paper No.18. Washington: World Bank.

Mearns, R. 1996. "When Livestock are Good for the Enviromnent." IDS Working Paper No.45. Brighton: Institute of Development Studies.

Nelson, R. 1990. "Dryland Management: The 'Desertification' Problem." Technical Paper No.116. Washington: World Bank.

Oldeman, L.R., R.T.A. Hakkeling, and W.G. Sombroek. 1990. *World Map of the Status of Human-Induced Soil Degradation: an Explanatory Note.* Revised Second Edition. Wageningen: International Soil Reference and Information Centre.

Ottichilo, W.K. 1996. "Wildlife-Livestock Interactions in Kenya." Paper presented at the Seminar on Balancing Livestock Production and the Environment, Washington, 27-28 September 1996.

Pagiola, S. 1993. "Soil Conservation and the Sustainability of Agricultural Production." PhD Dissertation, Stanford University.

Pagiola, S. 1994. "Soil Conservation in a Semi-Arid Region of Kenya: Rates of Return and Adoption by Farmers." In T.L. Napier, S.M. Camboni, and S.A. El-Swaify (eds), *Adopting Conservation on the Farm.* Ankeny: Soil and Water Conservation Society.

Pagiola, S. 1995. "The Effect of Subsistence Requirements on Sustainable Land Use Practices." Paper presented at the Annual

Meetings of the American Agricultural Economics Association. Indianapolis, 6-9 August 1995.

Pagiola, S., and M. Bendaoud. 1994. "Long-run Economic Effects of Erosion on Wheat Production in a Semi-Arid Region of Morocco: A Simulation Analysis." Agricultural Economics Staff Paper No.AE 95-12. Washington State University.

Pagiola, S., and J.A. Dixon. 1997. "Land Degradation Problems in El Salvador." In *El Salvador: Rural Development Study*. Report No.16253-ES. Washington: World Bank.

Pagiola, S., J. Kellenberg, L. Vidaeus, and J. Srivastava. 1997. *Mainstreaming Biodiversity in Agricultural Development: Toward Good Practice*. Environment Paper Number 15. Washington: World Bank.

Parton, W.J., G. McKeown, V. Kirchner, and D. Ojima. 1992. *CENTURY Users' Manual*. Fort Collins: Colorado State University.

Pieri, C. 1992. *Fertility of Soils: A Future for Farming in the West African Savannah*. New York: Springer-Verlag.

Pieri, C., J. Dumanski, A. Hamblin, and A. Young. 1995. "Land Quality Indicators." Discussion paper No.315. Washington: World Bank.

Pratt, D.J., F. Le Gall, and C. de Haan. 1997. "Investing in Pastoralism: Sustainable Natural Resource Use in Arid Africa and the Middle East." Technical Paper No.365. Washington: World Bank.

Repetto, R., and W. Cruz. 1991. *Accounts Overdue: Natural Resource Depreciation in Costa Rica*. Washington: World Resources Institute.

Sanchez, P.A., R.J. Buresh, and R.R.B. Leakey. 1996a. "Trees, Soil and Food Security." Paper presented at the Discussion Meeting on Land Resources: on the Edge of the Malthusian Precipice?, London, 5 December, 1996.

Sanchez, P.A., A.-M. Izac, I. Valencia, and C. Pieri. 1996b. "Soil Fertility Replenishment in Africa: A Concept Note." Paper presented at the Workshop on Achieving Greater Impact from Research Investments in Africa, Addis Ababa, 26-30 September 1996.

Sanchez, P.A., R.J. Buresh, F.R. Mwesiga, A. Uzo Mokwunye, C.G. Ndiritu, K.D. Shepherd, M.J.

Soule, and P.L. Woomer. 1997. "Soil Fertility Replenishment in Africa: An Investment in Natural Resource Capital." Paper presented at the Workshop on Soil Fertility Replenishment in Africa, Nairobi, 3-9 June 1997.

Schiff, M. and A. Valdés. 1992. *The Political Economy of Agricultural Pricing Policy* Baltimore: Johns Hopkins University Press.

Schroeder, P. 1993. "Agroforestry Systems: Integrated Land Use to Store and Conserve Carbon." *Climate Research*, Vol.3, pp.53-60.

Scoones, I. 1995. "New Directions in Pastoral Development in Africa." In I. Scoones (ed.), *Living with Uncertainty*. London: IT Publications.

Sombroek, W.G., F.O. Nachtergaele, and A. Hebel. 1993. "Amounts, Dynamics and Sequestering of Carbon in Tropical and Subtropical Soils." *Ambio*, Vol.22 No.7, pp.417-426.

Srivastava, J., N.J.H. Smith, and D. Forno. 1996. "Biodiversity and Agriculture: Implications for Conservation and Development." Technical Paper No.321. Washington: World Bank.

Steinfeld, H., C. de Haan, and H. Blackburn. 1997. *Livestock-Environment Interactions: Issues and Options*. Brussels: European Commission Directorate-General for Development.

Stoorvogel, J.J., and E.M.A. Smaling, 1990. "Assessment of Soil Nutrient Depletion in Sub-Saharan Africa: 1983-2000." Report No.28. Wageningen: Winand Staring Centre.

Thiollay, J.-M. 1995. "The Role of Traditional Agroforests in the Conservation of rain Forest Bird Diversity in Sumatra." *Conservation Biology*, Vol.9 No.2, pp.335-353.

Tiffen, M., M. Mortimore, and F. Gichuki. 1994. *More People, Less Erosion: Environmental Recovery in Kenya*. Chichester: John Wiley.

Tinker, P.B., J.S.I. Ingram, and S. Struwe. 1996. "Effects of Slash-and-Burn Agriculture and Deforestation on Climate Change." *Agriculture, Ecosystems and Environment*, Vol.58 No.1, pp.13-22.

Tucker, C.J., H.E. Dregne, and W.W. Newcomb. 1991. "Expansion and Contraction of the Sahara Desert." *Science*, No.253, pp.299ff.

Umali, D.L. 1993. "Irrigation-Induced Salinity: A Growing Problem for Development and the Environment." Technical Paper No.215. Washington: World Bank.

Unruh, J.D., R.A. Houghton, and P.A. Lefebvre, 1993. "Carbon Storage in Agroforestry: An Estimate for Sub-Saharan Africa." *Climate Research*, Vol.3, pp.39-52.

Van Orsdol, K.G., I.G. Girsback, and J.K Armstrong. 1993. "Greenhouse Gas Abatement Thought Non-Forest Biomass Production: Allocating Costs to Global and Domestic Objectives." Environment Working Paper No.59. Washington: World Bank.

Walling, D.E. 1988. "Measuring Sediment Yield from River Basins." In R. Lal (ed). *Soil Erosion Research Methods*. Ankeny: Soil and Water Conservation Society.

Williams, J.R., K.G. Renard, and P.T. Dyke. 1983. "EPIC: A New Method for Assessing Erosion's Effect on Soil Productivity." *Journal of Soil and Water Conservation*, Vol.38 No.5, pp.381-386.

Wisniewski, J., R.K. Dixon, J.D. Kinsman, R.N. Sampson, and A.E. Lugo. 1993. "Carbon Dioxide Sequestration in Terrestrial Ecosystems." *Climate Research*, Vol.3, pp.1-5.

Woomer, P. and M.J. Swift (eds). 1994. *The Biological Management of Tropical Soil Fertility*. New York: John Wiley & Sons.

World Bank. 1992. *World Development Report 1992: Development and the Environment*. New York: Oxford University Press for the World Bank.

World Bank. 1996. *Desertification: Implementing the Convention*. Second Edition. Washington: World Bank.

World Bank. 1997. *Guidelines for Climate Change Global Overlays*. Environment Department Paper No.47. Washington: World Bank.

World Bank. 1998. *New Opprtunities for Development: The Desertification Convention*. Washington: World Bank.

World Resources Institute. 1992. *World Resources 1992-93: Toward Sustainable Development*. New York: Oxford University Press.

Young, A. 1989. *Agroforestry for Soil Conservation*. Wallingford: CAB International.